With best wishes on your birthday

to Stewart

from M...

Au...

HAMLYN BIRD BEHAVIOUR GUIDES

SEABIRDS

ROB HUME

BRUCE PEARSON

HAMLYN

TITLE PAGE ILLUSTRATION *Here a group of Gannets has found a shoal of fish – Mackerel perhaps – and lets fly with a series of fast, near-vertical dives with great power. The Arctic Terns will be after much small prey, barely submerging compared with the deep dive of the plunging Gannet.*

First published in 1993 by Hamlyn Limited,
an imprint of Reed Consumer Books Ltd
Michelin House, 81 Fulham Road, London SW3 6RB
and Auckland, Melbourne, Singapore and Toronto

ISBN 0 600 57951 4

A CIP catalogue record for this book is available from the British Library

Page design by Jessica Caws
Map by Louise Griffiths

Printed in Hong Kong

CONTENTS

INTRODUCTION

One or two days before I began the job of writing this book in earnest, I was at Bempton Cliffs on the North Humberside coast, running through this introductory chapter in my mind.

It was a rare day of clear blue sky, with a fresh offshore wind and the poor, abused North Sea was as blue as ever it could be. Maybe it was not quite the translucent azure of the seas off north-west Scotland, which I believe are the finest I have ever seen anywhere, but it was doing its best. Many of the auks had already left the cliffs, the adult male Guillemots, one evening, having sat beside their chicks, calling and bobbing their heads until the half-grown, flightless young found the courage to leap from the ledges into the great unknown. It was, however, a good day to see Puffins, so close that I could see the tiny black pupils in their dark blue-brown eyes, and notice that the blueness or greyness of the eye was unlike the darkest, liquid pools of brown of the Guillemots' eyes.

Gannets had their chicks still on the ledges, many like children's cuddly toys, big, bulky and all fluffy white like the purest sheepskin. Others were already fully feathered, except perhaps for a ruff of white down around the neck. We read about synchronization of breeding at seabird colonies, especially Gannet colonies, but there were also a few pairs with chicks so small that they could not have been many days old, and quite obviously the span of breeding activity at this growing colony was wide.

The Gannets gave their curiously mechanical, deep-throated calls, while Guillemots whirred and growled and Kittiwakes sometimes burst into noisy choruses of ''s getting-late' (or 'kittiwake'). The sea below was dotted with black Puffins and lines of brown Guillemots, looking contrastingly pale in the sunshine. Fulmars lazily wheeled past, while Razorbills crouched in hollows near the top of the cliff face.

This was an all too rare 'fix' for me, the essential injection of my favourite kind of birdwatching. I love wild places. I can stay all day watching big birds of prey soaring about mountain gorges, or spend hours in a Norfolk wood waiting for a glimpse of a rare autumn warbler, or be excited by the sheer perfection of winter wildfowl on a sunny day when their colours shine and the dense flocks create patterns across a marsh, but it is seabirds that grab me most.

Perhaps it is living, now, in Bedfordshire, as far away as I can be from a decent rocky coast, that adds an element of rarity, of the unattainable, to all the other undoubted qualities of ocean-going birds that breed on coastal cliffs. Whatever it is – and I admit to a considerable dose of nostalgia for days gone by, summer holidays spent in Scotland watching the seabirds on Faraid or Dunnet Head and Handa Island or passing Ardnamurchan Point – it is a strong, magnetic attraction. Sometimes the need is so great that it almost hurts, and eventually I have to get back to the sea again, to catch up on my Guillemots, the Fulmars, the crying Kittiwakes and the impeccable Gannets.

Seabirds have always had this attraction for people. Some are immune to it and carry on watching Coots and Tufted Ducks or Pheasants and partridges for preference, or study their beloved tits or warblers. Others are hooked for life, and such great naturalists as James Fisher and Ronald Lockley would surely have given any number of warblers for a good petrel. Not many can fail to be thrilled by the sight and sound and smell (or, I would prefer to say, scent) of a big seabird colony, and the setting adds immeasurably to the appeal.

Flying Roseate Terns have quite fast wingbeats. They look curiously like Common Terns as white as Sandwich Terns, but with an action reminiscent of Little Terns!

The Herring Gull (left) and Lesser Black-backed Gull are sufficiently distinct in western Europe to present few problems, but the Herring Gull complex farther east becomes so confusing that it may be difficult to draw a line between 'Herring Gulls', 'Yellow-legged Gulls' and 'Lesser Black-backed Gulls' (let alone Iceland and Thayer's Gulls in the Nearctic).

It is so with many birds, the world over. No Golden Eagle looks as good as a Scottish highlands one, against a splendid backdrop of moor and crag and barren mountain top. No flock of Wigeons looks better than that which regularly grazes the wet grassy marsh beside the Dyfi in Mid Wales, where the winter light brings out an unsurpassed richness of colour in the surrounding hills. There can scarcely be a better place for a hunting Hobby than over some boggy New Forest pool in Hampshire. With seabirds, however, there the settings are so full of mystery and meaning, of folklore and legend, of danger and spine-tingling thrills.

There are terns on the rocky islets of Scilly where many a ship has foundered, its cargo plundered by the wreckers of past centuries. Shearwaters come to land on the darkest of nights on the upper slopes of the mountains of Rhum, set in the gleaming sea of the Hebrides, where people still tell tales in Gaelic at the fireside. I have counted auks and Fulmars on the mighty cliffs of Clare Island off the shores of Co. Mayo, among the friendliest people and on one of the most evocative coasts of Ireland, looking down from near-impassable sheep tracks on to dizzy cliffs and a crashing ocean swell. I have watched the Gannets on the rocks of Grassholm, far off the coast of Dyfed in the stormy Irish Sea, and the Puffins on the red sandstone and clumps of flowering thrift of Skokholm and Skomer. On Fetlar, home of the rare and majestic Snowy Owl, I have seen the Manx Shearwaters gather offshore in the Simmer Dim, the half-light that passes for night-time in Shetland summers. I have listened to moaning, musical fog warnings blasting out on misty days from cliffbound Grampian shores, while Fulmars have sped by at head-height and barely a wing's width away. I well remember my first Gannet, passing the lighthouse of Ardnamurchan Point. There is nothing to beat them. Seabirds, the coast and the ever-mysterious sea weave a potent magic.

Breeding seabirds are not, perhaps, of the stuff that makes the migration enthusiast, the keen birder or the dedicated twitcher feel a quickening of the pulse and an urge to get the car out and go. They prefer the autumn movements of shearwaters and petrels or the spring migrations of skuas past long headlands that allow a brief look out into the world of the sea far offshore. An onshore gale in autumn brings with it the possibility of rare Sabine's Gulls, Leach's Storm-petrels, perhaps Cory's Shearwaters or Long-tailed Skuas.

Seawatching, the art of staring out to sea and identifying the seabirds as they pass by, has become ever more popular in recent years. Specialist guide books have dealt with all the birds of the world's oceans, and a surprising number of them have been found and reliably identified thousands of miles from their normal range. An American Least Tern has taken up a regular summer home on the English south coast, while an African Lesser Crested Tern has reared hybrid young with a Sandwich Tern for a mate in the Farne Islands, the very place where an Aleutian Tern from the Pacific created the greatest stir of all until an Ancient Murrelet, far from its North Pacific Ocean origins, decided to spend several successive summers on Lundy Island off north Devon.

Even these were at seabird colonies, but the expectation of a rarity on migration has a different appeal. It is an equally strong one,

and seabirds exert it the length and breadth of the United Kingdom, with magnificent hotspots in the seawatching world from St Ives and Porthgwarra in Cornwall, Cape Clear Island in Eire and Beachy Head in East Sussex to Flamborough Head, Hartlepool, Hilbre Island and far-flung Balranald in the north and west. All of these, however, depend on the weather, especially the wind. A bad day at Portland Bill or Cley, calm weather at Filey Brigg or an offshore wind at West Kirby can deliver the dreariest, most tedious bird-watching – or non-birdwatching – of all. As Bill Oddie once wrote, such days breed experts in the art of hitting tin cans with stones, but give precious little to write home about in the way of good birds.

However you like your seabirds – and I dare to presume that you do – they fall neatly into particular groups. They feed, fly, breed and even migrate in distinctive ways. Certain groups share particular plumage features, or structural adaptations. They can be organized, rationalized into clusters of more or less similar species for ease of comparison and distinction. This book does that, and in particular examines the behaviour of our seabirds in the sense that we can easily see it for ourselves: the way they fly, the way they plunge for fish or dive from the surface or scavenge at tips. It is a guide to the seabirds of western Europe and the North Atlantic, not a strict, simple identification-by-appearance book, but a behaviour guide, which aims to supplement the available field guides by devoting more space to what birds do and, consequently, what they look like. Above all, it will attempt to convey something of the unique appeal of this group of birds, open to all who can appreciate a good thing when they see it. Seabirds are wonderful.

Heavier and stronger than most terns, Sandwich Terns dive from a height and enter the water with a loud splash. They take larger Sandeels from a greater depth than their smaller relatives can manage.

Classification and Distribution

Birds of the oceans, many breeding on remote islands or sheer cliffs, some coming to land only after several years' immaturity at sea and only at night, include the most mysterious and romantic of birds.

Evolution has fitted them for life in an environment which, superficially, seems relatively uniform: the sea. Yet a deep examination of their life-styles reveals just how varied they are and how their environment presents a huge range of opportunities to exploit and hazards to survive. Their classification reflects their diversity and similarities, while their form and function – how they look and how they live – shows their superlative adaptations to an ocean-going existence.

A checklist of all the birds of the world, beginning with the oldest, most primitive groups (species clumped together in closely related families, in turn grouped into orders), moves from ostriches and emus

Red-billed Tropicbirds are spectacular plunge-divers, almost like Gannets. Magnificent Frigatebirds soar above, hoping to steal a meal.

through tinamous and quickly finds the first seabirds: the penguins. These flightless and famous birds of the southern hemisphere form the seventh order in the world list, followed quickly by divers and grebes (both of which have some claim to be seabirds, although they breed beside fresh water), the tubenoses (which include such classic seabirds as albatrosses and storm-petrels) and the large, mixed pelican group, among which tropicbirds, gannets, boobies, many cormorants and the frigatebirds are true birds of the seas. Then there is a large gap, filled by such groups as herons, wildfowl and birds of prey, before the next order containing seabirds, these of a rather different nature.

This is a large and varied order, the Charadriiformes: the waders, skuas, gulls, terns, skimmers and auks. Waders are not, in this sense, seabirds (though many live beside the sea, and the phalaropes spend much of the year upon it), but the rest of this order is largely an ocean-going one (although many gulls and some terns live beside fresh water far inland). The list, by now, has included sixteen orders and 63 families of birds. In the remaining eleven orders and 110 families of the complete world bird list, not a seabird is to be found.

Penguins need not concern the northern-hemisphere or North-Atlantic birder further. Sadly, the eighteen species come no farther north than the Galapagos and, in the Atlantic, not much beyond the southern tip of Africa (and then only the Jackass). Albatrosses are

almost as easily dismissed, being fourteen species of the southern oceans (and Pacific tropics), where winds roar around the earth free of any hindrance. A few species, however, do move farther north, and an occasional individual breaks through the Doldrums (where winds are light and flight difficult for the sailplane albatross) and reaches the North Atlantic. So, of the albatross, we shall hear a little more later.

Petrels and shearwaters, however, have important representatives in the North Atlantic, including such a variety as the Fulmar, Leach's Storm-petrel and Manx Shearwater. The ninety-odd species are closely related to albatrosses, but none reaches the 3-m-plus wingspans of those elegant giants. Indeed, the European Storm-petrel is among the tiniest seabirds. Diving-petrels, which are close to the true petrels but appear remarkably similar to the northern auks, are four species of the far south, never recorded in the North Atlantic.

Tropicbirds (three variable species) are tropical creatures with no more than the odd vagrant getting into the North Atlantic proper, so they can be dispensed with here (save that they include some of the most beautiful birds on earth), along with the six boobies. The latter are also warm-water birds, getting up to the tip of the Red Sea and, rarely, alongside the eastern coasts of the USA, but not properly penetrating the North Atlantic range with which this book is concerned.

Gannets include three very closely related species, but it is not certain how often one (the Cape Gannet) reaches north of the Equator, while another (the Australasian Gannet) is irrelevant to us. That does, however, leave one of the most wonderful seabirds, the Gannet, or Northern Gannet, which is very much part of our coastal scene.

We see but three cormorants, unless we add the elegant, rather rare Olivaceous Cormorant of southern USA and Caribbean coasts. Elsewhere, there are some thirty extra species and many are striking, beautiful birds. Only in the Gulf of Mexico or at the tip of Florida are frigatebirds more than a wild dream of a rarity-conscious seawatcher: most of the five kinds are Indian Ocean and/or Pacific species.

We do better for skuas, with four regular and a fifth very rare visitor of the seven species. Gulls are familiar and important, as they are almost everywhere (a visitor to tropical island groups such as the Seychelles will marvel at their absence). Of the world's 48 gull species, the North Atlantic is home from time to time to some nineteen kinds. Terns include a number of freshwater and marshland species, but they come to sea in winter. There are around 43 species, with fourteen regular and one or two rare in the North Atlantic, although but one skimmer can be mentioned and that only for the eastern USA.

Auks are somewhat penguin-like, but smaller and capable of flight (although the recently extinct Great Auk was flightless). They occur in great numbers, but only six of a world list of 22 are regular, while the others are so rare as to be best left alone here.

THE OCEANS

Like the landmasses of the earth, the oceans have undergone enormous changes in shape, size and distribution over time as early continents have split and drifted apart across the planet's surface. The continents can be thought of as a crust on a meringue, or the solidified skin on a bowl of custard, which breaks into discrete blocks which slide gently across the semiliquid material beneath. Unlike the land, the oceans remain interconnected: there is no isolated, landlocked sea of importance on a global scale, and no seabirds to compare with the seals of Lake Baikal or the dolphins of the Amazon in terms of total isolation.

As we now see them, the oceans form five major bodies: the Arctic and Antarctic seas (the former frozen in the centre, the latter ringing a landmass), two north–south oceans – the Atlantic and the broader Pacific – which connect the two polar seas and one large, broad ocean coming from the far south only to be landlocked on its northern side, the Indian Ocean. There are, of course, many irregularities and some seas that are to a greater or lesser extent isolated from the main oceanic mass, the Caribbean, Mediterranean and North Seas being obvious examples.

The oceans are far from uniform in their suitability for birds. Where cold currents meet warm seas fish thrive, attracting birds in great numbers. Where tropical islands are surrounded by warm seas with relatively few fish, the birds must behave differently if they are to get a good living: different species, with different life-styles, have evolved here. In stormy oceans, where wind is almost constant, great birds can travel huge distances at speed with little expenditure of energy and the long-winged, sailplane gliders such as albatrosses and larger shearwaters perform to perfection. Where the wind is erratic, often almost absent, such birds cannot survive. They are unable to progress through calm air without tiring themselves out, beating those long wings which are meant to stay outstretched and immobile. Becalmed albatrosses are as helpless as sailing ships in still air or hang-gliders lost over a flat plain. Again, birds have adapted along different lines to meet the challenges presented to them in regions of little wind.

In the south, the prevailing wind is westerly, sweeping around the subpolar regions in an endless circle of gales. The roaring forties and infamous gales farther south, epitomized by the fearsome storms which are so often encountered around Cape Horn, create the most

The Black-browed Albatross is one of a group of relatively small albatrosses termed 'mollymawks', but in North Atlantic terms it is still a giant among seabirds. Its span of more than 2 m and the narrow wing shape allow the efficient exploitation of surface winds for long-distance travel.

testing conditions for mariners, but support a wealth of seabirds that use the winds to their advantage. How such small specks of life can beat us humans when it comes to life on the ocean wave!

Three high-pressure centres exist just to the north of this belt, one in each of the major oceans, creating a broadly anticlockwise spiral of winds in the southern parts of the Pacific, the Atlantic and the Indian Ocean. In the Indian Ocean and the western Pacific, the tilt of the earth and the lie of the land produce marked seasonal changes, creating the monsoon winds that blow from the north between November and April and from the south at other times. In the eastern Pacific and Atlantic, in a belt south of the equator, the south-east trade winds are prevalent all year, while to the north of the equator the north-east trades blow steadily: they meet in an area of little wind, the equatorial Doldrums, which create such a barrier to the riders of the wind from the south.

In the north, westerly winds cross the Atlantic and sweep north-west along the coast of Europe. There are frequent storms centred on the Gulf of Mexico and Caribbean which rage north-eastwards across the Atlantic, especially in autumn. Seabirds can usually cope, but sometimes a combination of onshore winds and lack of food drive them ashore or even far inland in large numbers in north-west Europe. These 'wrecks' are irregular and mysterious: not every such episode produces a seabird wreck, and not every seabird wreck involves the same species of seabird.

After a prolonged onshore gale, Leach's Storm-petrels may give up the unequal struggle and allow themselves to be drifted inland. They usually manage to find a large lake or reservoir, being seen in unlikely circumstances for a day or two before moving back to sea.

In north-west Europe the summers are mild, with long days and little darkness in the far north, while the winters are cold, dark, with only brief hours of daylight. Compared with the relative uniformity of the weather in the tropics and the stability of the oceans farther south, where the extreme warming and cooling of landmasses has less effect, the North Atlantic enjoys, or suffers, much more erratic, more extreme changes in weather, from season to season and within the less reliable seasons themselves. The long days allow birds to feed almost around the clock in summer, but they have to move south in winter. The fickle summer weather, however, not only affects the ability of the birds to find food but also has enormous influence on the food sources themselves: the fish and plankton that form the basis of so much seabird ecology.

The ocean currents reflect the winds, with a westerly circulation of cold water in the far south producing offshoots northwards along the west coasts of South America and Africa, each turning westwards close to the equator. These narrow injections of cold, nutrient-rich water into the warm tropical seas are immensely significant for the great concentrations of seabirds along the western coast of South America, particularly, and for the many that breed to the north but spend the winter off West Africa.

A broad equatorial current sweeps west from south-west Europe and North Africa to the Caribbean, from where a long, warm current crosses the Atlantic, heading north-east. This is known as the Gulf Stream, responsible for the mild winters of much of north-western Europe compared with the vicious cold of the Continental winters not far to the east. Cold currents, rich in food for fish and seabirds, come down from the Arctic to penetrate the most northerly limits of the Atlantic proper, and here seabirds prosper in ideal conditions. In parts of the North Sea, cold-water upwellings and the meeting of currents produce conditions that suit plankton and plankton-feeding fish to perfection, and the birds take full advantage.

On to these patterns must now be superimposed the influence of man. This is manifested in three main ways: pollution, which affects birds directly (particularly the spilling of oil) and through their food (many kinds of poisons become concentrated in fish, or reduce the number of fish in the sea); direct trapping of birds by fishing gear (nylon gill nets and the huge 'wall-of-death' drift nets, which are sometimes several tens of kilometres long, catch hundreds of thousands of diving birds annually); and overfishing, which may reduce the number of fish available for seabirds to exploit. In some areas, such as the northern North Sea and the Barents Sea, two or three of these factors may operate together, and seabird productivity and survival rates have plummeted in recent years. In a few places, such as the seas around Malta, man still exerts a limiting influence by killing seabirds for sport. Shooting birds from boats persists in the Mediterranean, but thankfully most such shooting has been surrendered in western Europe, if not always in the north.

Some seabird trends in Europe have been frightening and tragic. In the Shetlands, Arctic Terns, and even skuas and Kittiwakes, reared almost no young at all during the 1980s. In the Norwegian

Arctic Terns are small and light and cope with their huge migration flights by flying slowly and steadily, feeding as they go. In their breeding areas they depend on a supply of small fish, such as Sandeels, but are always likely to be robbed by an Arctic Skua (top right).

and Barents Seas, famous seabird colonies on splendid, remote islands crashed, their inhabitants drastically reduced in number after years of poor breeding success. Overfishing, particularly 'factory fishing' for animal foodstuffs and even power generation (by burning oil from Sandeels), was implicated in some areas, deaths of vast numbers of birds in fishing nets in others. In the main, however, the cause was not clear: there seemed to be several factors involved, and fishing was perhaps but the last straw (although it was clear that better controls were needed, and fast). The ecology of the little Sandeel, so important a fish in the lives of our seabirds, suddenly became all important.

Form and Function

Seabirds exhibit a remarkable variety of shapes and sizes. They tackle the demands of life on or beside the ocean, and make a living from what it has to offer, in their own different ways.

Their variations are closely connected with the food they eat and the means by which they obtain it. Some travel to catch fish that are themselves local in their distribution and found far from land, some plunge from the air while others dive from or from near the surface, relying on the speed of their dive to take them underwater, or using their wings to reach greater depths. Some forage for floating plankton and also need to travel; others stay close to their home shores.

We have already seen that **albatrosses** and larger **shearwaters** exploit the energy of winds and the air currents that develop over steep ocean waves in order to travel long distances with little waste of energy – indeed, with little use of energy. Long, slender wings are ideal for gliding flight, the length creating a large surface area to catch the air currents, while the narrowness gives a reduction in drag, allowing

ABOVE *Friction drags the wind to a lower speed immediately above the waves, but a little higher,the air is unhindered, so it moves faster. Birds take advantage of this by shearing upwards and gliding along in the high-speed air.* OPPOSITE *Shearwaters (like these Great Shearwaters) and many other seabirds use dynamic soaring, sweeping up into the wind, tilting over to breast the breeze and running with it, or across it, before the next upward sweep. In this way they progress quickly with little expenditure of energy.*

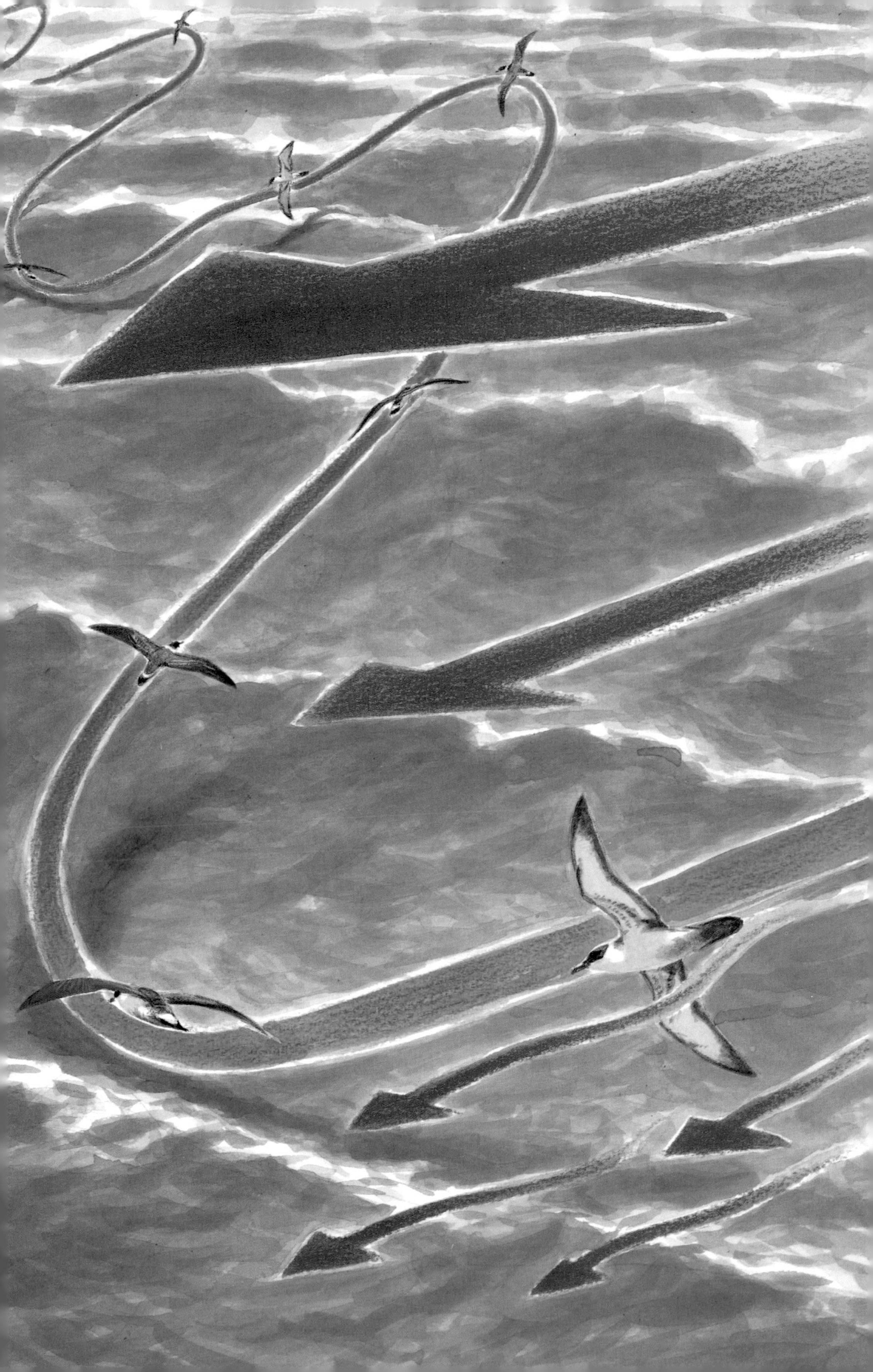

the bird to slide through the air at speed. The length is achieved by a very long 'arm', the bones of the inner wing (equivalent to humerus and forearm) being remarkably long, with an unusually large number of feathers, while the 'hand' is less elongated, giving rigidity at the wingtip. The price is reduced manoeuvrability and even an inability to move easily on land, to which all must resort when breeding.

An albatross or a **Great Shearwater** flies across the wind, shearing up on one wingtip to present its undersurface to the breeze, then turning downwind and losing height in a long, fast sweep. At the end of this downward glide it turns back upwards and into the wind, ready for the next downward, onward sweep. In this way, it makes fast and reliable progress downwind with few wingbeats, or none at all, converting the energy of the wind and gravitational force into forward motion. In the northern waters around Britain and Ireland, we tend to see the Fulmar performing like this most frequently, although the Manx Shearwater, Gannet and Kittiwake are equally adept at dynamic soaring in the right conditions.

Should the wind drop, however, the bird will struggle in calm air. The long wings become a hindrance, hard to beat energetically, with the small breast muscles having to force unwieldy limbs down in an effort to achieve worthwhile motion. This is why the Doldrums are a barrier to the southern albatrosses. **Manx Shearwaters**, much lighter, smaller and relatively short-winged, cope better with occasional still days and flat calm seas, although they too progress in a series of wingbeats followed by short, flat glides. **Little Shearwaters** around the Canaries and Azores have a more fluttery flight with several quicker wingbeats and shorter glides, parallel with the water. Fulmars and larger shearwaters, such as the **Cory's Shearwater** of Mediterranean and mid-Atlantic seas, progress with a notably heavy, uncomfortable look. Sunshine and flat seas make the bird fly with lumbering, heavyweight flaps and occasional turns, using one wingtip as a fulcrum as the whole body arcs over the top. A sudden squall gives enough wind to free it from this manual labour, and its upward arcs become high-rising sweeps clear of the water, with no wingbeats necessary and progress in a higher gear, swift and easy. Once the squall has gone through, perhaps after just a half-minute or minute, the bird subsides into its weighty, uncomfortable motion again.

Smaller petrels require strong flight for long-distance travel and to weather the strongest of storms at sea, and a lightness of touch to stoop down to pick tiny items of food from the surface. They have rather short, broad tails (some forked) to improve manoeuvrability in a tight situation, and quite short, often broad-tipped wings that give the benefits of dynamic flight in wind and over waves together with an ability to flutter and flap, enabling fast progression even when the air is still. They are so small and light, in any case, that flight is never such an effort as it sometimes is for their bigger, heavier relatives, and buoyancy is the abiding quality of their flight.

Leach's Storm-petrel is a terrific performer, with erratic twists and sideslips combined with shearwater-like glides, while the **European Storm-petrel** is a flutterer, often with a more swallow-like series of soft wingbeats and tilting glides together with sudden stalls and dips. Some petrels, like the two already mentioned, are characterized by different ratios of 'hand' to 'arm', which produce differences in flight as well as proportion that are so useful to the experienced seawatcher.

The **Gannet**, too, uses a number of flight actions. In a gale, it is a perfect glider, yet it can fly for hours with a steady, economical action, a series of strong beats between flat, or sinking, dives and banks. It has a long, tapered body, and a long, pointed tail to help it plunge for food, and rather long, narrow, pointed wings without the albatrosses' extreme proportions. When soaring around its breeding cliff, it makes subtle changes of wing shape, now extending them flat and straight, now flexing the joints, easing the tips inwards and angling the wrists and elbows to reduce their length. The outermost primaries curve and lift away from the rest, allowing air to slip through a slot at the wingtip to prevent a disastrous slow-speed stall. It is streamlined to minimize drag and maximize the ease with which it can penetrate the air. More an ocean-going glider than an allrounder, it can cope with most weather. Gannets at the colony are magnificent to watch.

Cormorants and **Shags** remain in coastal waters to exploit a more reliable, less seasonal, food source which they reach in a deep dive. Their flight is heavy but strong, not specialized, although the Shag tends to fly fast and low while the Cormorant rises high in the air and forms goose-like chevrons and 'V's when in flocks. They do not have the soaring abilities of their migrant relatives, the pelicans, nor do they dive headlong like Brown Pelicans, related gannets and tropicbirds. Nevertheless, they may surprise one by their antics at a colony or winter roost: they regularly arrive from a great height, whiffling down in sideslips and barely controlled headlong tumbles, before circling and heading neatly in to their chosen perch. The landing may be heavy, but the approach is better than they are given credit for.

Frigatebirds watch for other seabirds feeding. They soar at height to get a good view from a stable base and then accelerate to get in fast and steal their food. They need power for speed, and their long, angular wings propel them forward swiftly while also picking up the least breath of air to help them along. Once they reach their victim, they chase it fast and furiously, matching its turns and twists. Again, their long, flexible wings and long, deeply forked tails provide the control surfaces needed for brilliant, tightly controlled manoeuvres that belie their great size. Long, hooked bills are the last of their special attributes, just right for hoiking a dropped fish out of the air.

There is no one solution to the problems of life at sea: Manx Shearwaters (top three) have long wings and glide low over the waves; Leach's Storm-petrels have erratic but powerful flight; the Storm-petrel (bottom right) looks feeble but still survives a life over the open ocean.

Skuas do much the same, but with less extreme adaptations. Nevertheless, they have an easy, rhythmic flight oozing power and authority, and they can migrate long distances or take on a gull or tern in close combat. **Gulls** lack the skuas' pace and dynamism, but have an all-round ability in the air that few birds can beat. Their wings are long but quite broad, combining a large surface area with an efficient, low-drag configuration, often helped by a bowed or angled shape. Short, square tails are no match for the deeply cut, streamered tails of terns or the spiked tails of skuas, which add extra control in a dive or turn, but the gulls' wings, coupled with free use of their legs and webbed feet, give them an ability to turn on a sixpence.

Larger species are also expert gliders and make full use of ships' slipstreams, gliding along with ease at high speed until they pause to pick up scraps, after which they drive along in a powerful chase, with regular, deep wingbeats, in order to pick up the tow again. The differences in action of the various gulls, especially related to their size, are obvious to the experienced eye. A Common Gull is light and gentle compared with a Herring Gull, but in a flock of Black-headeds

it seems big and broad-winged. A young Mediterranean Gull has a plumage pattern not unlike that of a young Common, but is small and quick, stiff-winged and flappy compared with the fluency of a Common: it looks like a Common that flies like a Black-headed.

The tail streamers of a seagoing **tern** assist it to aim a headlong dive precisely at fast-moving underwater prey. The fractional twists and tilts in the dive, making the final adjustments in angle and position, come from the tail and the long, pointed wingtips, which act as sensitive ailerons in the plunge. Their long wings also give the terns strength and stamina in long-distance migrations, when they use an economical, rhythmic flight action unencumbered by the weight and clumsiness of the albatross or big shearwater in calm air. The lightest, airiest terns have a distinct up-and-down movement of the body in compensation for the movements of the wings.

Auks are quite different. If they move a long distance outside the breeding season, much of it will be swum. In summer they fly to and from their feeding grounds, but not far. Their flight is inefficient, because they are designed to perform underwater. Here they chase fish in fast, underwater 'flight', using their relatively small, short, but narrow wings in pursuit. Those same wings are not too good in the air: they are a compromise, and provide insufficient surface area to propel the bird aloft without constant, fast, almost whirring beats, their heavy bodies hanging below as if the effort is almost too great.

In the late summer and autumn, adult Common Terns moult their wing feathers and often show obvious signs of missing primaries (left) while the Arctic Tern (right) can be distinguished by the completeness of its wings, which are moulted only when migration is complete.

PLUMAGE AND MOULT

All birds must periodically change their feathers, in the process known as moult. This shedding and regrowth is not random. Each species has a defined, unchanging order in which the feathers are dropped. **Gulls** are the most easily studied, because they are large, easily seen and usually have grey, brown or black feathers which, when shed, reveal the white bases of other feathers beneath. Young, 'brown' gulls changing into adult grey and white reveal the contrast between old, dark feathers and new pale ones, so the progression of the moult can be observed on a living, free bird in a way rarely achieved with a small songbird, for example, without catching it.

In autumn, it is easy to see adult gulls with ragged-looking wings and long, irregular, whitish bands across the upper surfaces. This is because the flight feathers (primaries and secondaries) are being moulted, producing an irregularity in the trailing edge of each wing, and their coverts are also in moult, revealing white patches where darker feathers have fallen. The moult, for most gulls and terns, begins with the innermost primary feather, so that the first sign is a notch behind the bend of the wing. It proceeds over the head, body, wings and tail. The progress can be judged by the primary-feather moult: by the time the last, outermost primary is shed and replaced, the moult, which may take two months or more, is over. One or two primaries may be missing from each wing, along with two or three secondaries, while others are new feathers still only half-grown.There will be gaps and notches in the wing, and it is obvious that the bird is not performing to its full potential. Flight is frequently quick and jerky.

Moult is strongly related to migration. **Glaucous Gulls** may begin their moult in late spring, then suspend it during the breeding season and complete it in their winter quarters. **Lesser Black-backed Gulls** from Scandinavia have slightly different moult schedules from British ones because they migrate further. **Arctic Terns** moult their flight feathers late in the year, after migrating south, but **Common Terns** moult their inner flight feathers twice a year, once while still on their breeding grounds, leaving the outer feathers until the winter after they have migrated (so they migrate with a full complement of flight feathers, letting them fly most efficiently when they need to).

In general, immature gulls moult earlier than adults, and adults which have failed in a breeding attempt moult earlier than those that rear young. The point is that feeding a family of hungry youngsters, and flying hundreds or thousands of kilometres on migration both place great demands on birds, and to be at their best they need a complete set of good feathers. Moult is delayed until after the family is independent (but, if the bird is too young to breed or has failed, it can moult earlier and less hurriedly), and is avoided during migration. Different species have evolved different schedules and, in a few cases, this is useful to the birdwatcher who wishes to identify them.

In the **auks**, moult takes a different form, in that it is completed after breeding but quite quickly. Because the birds no longer need to fly back to their colonies with fish, they can settle on the sea where there is a good supply of food and moult all their flight feathers in one go. They become literally unable to fly for a while.

A danger is that onshore winds blowing unseasonably strongly and consistently can drive exhausted, flightless auks ashore. If rough seas prevent them feeding, they tire easily and cannot resist the remorseless drive towards the dangerous beaches. Occasionally, large numbers are brought ashore in 'wrecks' in this way. Only rarely, however, are Guillemots and Razorbills taken far inland, in the way young Manx Shearwaters regularly and Leach's Storm-petrels sometimes are.

FOOD AND FEEDING

The food that a seabird eats rules its whole life. The bird is already pre-adapted to its particular food, the diet that its ancestors have eaten over thousands of generations. It can do little else but behave in a pre-programmed way determined in the egg, and use its mental and physical attributes to find, catch and eat the kind of food to which it is suited. Seabirds have little intuitive or intelligent capability to turn to other ways of life or feeding techniques than the one they were born with, although experience hones their skill and older birds may perform better than young ones, which take time to perfect the difficult job of surviving. It is food that determines their movements, their general shape and detailed structure and much of their behaviour.

Gannets eat large, muscular, fast-swimming fish that are found not far below the surface of the sea. To get to them they must dive, but to see them they need a good, extensive view. So they do not swim about and dive from the surface; instead, they fly high and look from mid-air, using a better vantage point with the chance of a more instant reaction. When it spots a fish, the Gannet dives. It hits the water headlong, and hard, sending up a jet of water and spray that can be seen several kilometres away. Then it submerges, grasps the fish in its bill and swallows it, before flying up again and moving on.

To do this, the Gannet has certain attributes that allow such a hunting method. It must be a strong flier, able to glide and soar high up, while looking downwards at the sea, and with the ability to halt forward progress by suddenly rising, even turning, while not taking its eyes off its intended prey. It has to be able to twist and position itself in the dive, so that its strike is accurate and efficient. So a long tail, pointed in the centre, is useful as a control surface.

It must judge the position and depth of the fish very accurately. To help it, it has forward-facing eyes that give excellent binocular vision, so that it can focus on the fish with both eyes at once, using the properties of parallax to assess the location of its prey. With one eye, judgement of distance is difficult; two make it easier. The lines of sight of both eyes converge at the fish, allowing the Gannet's brain to judge the distance of the crossing point. It works in the same way as the method used by the wartime dambusters, who judged the precise height over water at which to drop their bouncing bombs by

Strong, muscular fish like these Pollack are the prey of Gannets, whose binocular vision enables them to judge their dives to perfection.

shining lights from the front and rear of the aircraft and descending until the two beams met – simple triangulation. We do the same but, like the Gannet, do not even think about it. It is easily demonstrated. Look at an object a couple of feet away – the corner of a book on a shelf, for example, or the end of the arm of your chair. Touch it with a fingertip. Now try reaching out to another book, or the other arm of the chair, but this time with one eye closed. It will not be so easy, not so accurately, yet unconsciously, achieved.

The Gannet is protected from the impact of the dive by air-filled sacs under the skin of its head and neck and spongy bone at the base of the bill. Special membranes protect its eyes, and its nostrils are arranged to prevent water penetrating its nasal passages.

Like so many other seabirds, the Gannet is a model of adaptation, and it is mainly its food that has determined the way in which it has evolved. There are many more examples of such beautiful evolution to a specialist way of life centred on getting enough to eat.

The **Cormorant**, a relative of the Gannet, feeds on slower-moving fish found at the bottom of the sea. It does not spot them from the air, but dives underwater from the surface and swims until it finds suitable prey. Its broad, hugely webbed feet provide underwater propulsion (the Gannet has similar feet, but its motion underwater comes from the speed of its entry). The Cormorant reaches out and grabs a fish in its bill, which is long and strong with a sharp, hook at the tip to hold on to a struggling fish. To help the catch, the Cormorant has a longer, more flexible neck than the Gannet, of a kind that would be unsuitable for the high-speed impact of a 30-m dive.

The relationship of Gannet, Cormorant and pelican can be seen by looking at a few photographs. Superficially they look different, but the structure of the bill, although the final shapes are so unalike, and the webbing of the feet are the same; and the bone structure and feathering of the spread wings have much in common. It is easy to see that these birds are close to each other in the order Pelecaniformes.

Small **storm-petrels** that feed on surface-dwelling plankton need an altogether different technique from that of Gannet and Cormorant. They must be able to flit along the surface of the sea, head turned downwards, to pick food from the water with their bills. They need accurate reflex movements to do so efficiently and have a flight pattern that is remarkably buoyant with instant manoeuvrability. They slide and glide, flutter and dance, turn back on their track and even patter on the surface with outstretched feet, as they exploit the tiny items of assorted flotsam that make up their regular diet.

Some **terns** are specialist plunge-divers, with forked tails and long wings to help them hover and dive, long bills to grasp fish, and tiny legs because they do little on land but stand still. **Gulls**, in contrast, are mostly generalists, opportunists that have strong bills but no special adaptations for plunge-diving, long wings but broad, rounded tails and longer, stronger legs that allow much greater mobility on land, where some species feed as much as or more than they do at sea. They walk about on grassy swards looking for worms, or forage on beaches turning over the tide wrack or exploring rock pools for crabs. They can also fly long distances, however, and pick up items disturbed by a ship's propellers or brought up in a trawler's net.

The skimmers, including the **Black Skimmer** of North American coasts, are surface-feeders with a unique and remarkable adaptation. The long bill is deep and strong, but, while the upper mandible is normal, rather tern-like, the lower one is drawn out into a long, laterally flattened, blade-like tip which extends well beyond the upper mandible. This is dragged through the water as the bird flies, until it touches a fish or some other item of prey. The bird then closes its bill on its victim and slides it up towards its mouth. A noteworthy trait of the Black Skimmer is that it feeds at night, cutting a trail across the water that becomes lit up by disturbed, phosphorescent plankton. The light attracts fish, and the skimmer, clever bird, turns back on its tracks and skims up the inquisitive fish on its return journey.

Food influences the migrations of seabirds. If they catch fish that move seasonally (whether horizontally or descending too deep), they, too, have to move, perhaps to entirely different areas where new supplies can be exploited. So the long-distance flying capabilities of shearwaters, petrels and terns, for example, ultimately came about as a response to the need to move to find food.

Most seabirds eat fish. At the launch of Peter Harrison's trend-setting book *Seabirds*, he signed my copy with the message 'Real birds eat fish!'. Of 270 species of seabirds, some 138 eat mainly fish and another 62 eat quite a lot of fish. The fish-eaters include a broad variety, from albatrosses to cormorants, from shearwaters to auks.

At least forty species eat Crustacea more than anything else, and at least 138 species eat a good proportion of crustaceans in their diet – shrimps and shrimp-like creatures, large and small. Petrels, some shearwaters and the Fulmar are regular crustacean-gobblers, but several, especially the Fulmar, are quite able to live on other things, including fish and jellyfish. Squids, or cephalopods, form an important part of the diet of many seabirds, although only about fifteen rely mostly on them and these are largely southern species.

Dead fish and such floating carrion also provide birds with food. The gulls are great ones for picking bits and pieces of offal and upturned fish from the surface, but the **Fulmar** has perhaps been most closely associated with dead meat in northern seas. It has increased in numbers and range in the nineteenth and twentieth centuries, and this rise in its fortunes has been linked to the increase in whaling and, later, trawler activity. Fulmars certainly gather in

great numbers around factory ships, feeding on the offal that spills overboard or on the undersized fish that are thrown back because of commercial pressures or legislation to conserve stocks (but which are already likely to die). Indeed, around a trawler working over its catch, Fulmars and gulls compete for the rich pickings, while fending off the attentions of skuas and keeping an eye out for diving Gannets which spear past their heads. The eager gatherings can be exciting and dramatic, drawing in great numbers of birds from an apparently all but empty sea. Like vultures over the African plains, the seabirds are in 'extended flocks', widely dispersed but eyeing each other carefully. When one makes a determined move towards some food source, it is followed by another, which in turn is followed by another and so on, until a whole host gathers at a central point where none was visible before.

It is food, then, that generates the diverse adaptations of seabirds in terms of physical attributes (including plumage colours) and behaviour. It also determines their distributions, both in the breeding season, when they have to be within easy flying reach of a reliable source of food so that they can return to feed incubating mates and/or chicks, and in the non-breeding periods, when they are free to roam far from land but still need to eat. It is the availability or lack of food that explains why some sea coasts are awash with seabirds while other cliffs and islands have none. It is also responsible for occasional 'wrecks' of starved seabirds no longer able to resist onshore winds and which are washed ashore, and for periodic catastrophic breeding seasons when young birds starve in their thousands before they are old enough to fly.

FEEDING BEHAVIOUR

When seabirds are feeding, they fly, dive or swim in ways that are characteristic of their family, or in some cases even highly distinctive of individual species. Like any set of species that live in the same area, overlapping in time and space, they must differ in some way so that they can survive without constant competition. It may seem that similar seabirds living practically side by side – Puffin, Guillemot, Black Guillemot and Razorbill, for instance – must be

OPPOSITE *The increased amount of fish offal spilled overboard from trawlers in the early decades of this century has been quoted as one reason, perhaps the main one, why fulmars increased their numbers and spread their range so dramatically. Fulmars, gulls and Gannets crowd around trawlers, seeking injured fish brought to the surface, offal from processing at sea and undersized fish that are thrown back, unlikely to survive.*

pursuing the same food in the same way. It is possible, of course, for such circumstances to arise, if there is such an abundance of food that all the birds can take their share without fear of conflict in a great free-for-all, but the norm is for each species to be doing something rather different from the others, so that they can all get along quite well without each doing the other out of a living.

This diversity of adaptation, to take advantage of a whole series of opportunities that may present themselves, is quite usual among birds of many situations. Woodland species show it well. Great Spotted Woodpeckers take beetle larvae from deep within old wood; Treecreepers take insects and spiders from the surface of cracked, indented bark. Great Tits feed around the larger branches, Blue Tits on the side branches, while the tiny, lightweight Coal Tits can hang right out on the thinnest, springiest outer twigs that even Blue Tits with their minute weight bend down.

There may be a means of excluding competition within the same species. Male Sparrowhawks are small and light, catching tits and sparrows; females are larger, stronger and capable of catching thrushes and pigeons. In this way, the two can live together in a much smaller area than if they were both relying on the same species of prey: in effect, the density of Sparrowhawk pairs can be almost doubled.

It is the same with seabirds. Cormorants and Great Northern Divers are deep-divers, taking flatfish and sluggish prey in the dark depths, well below the fast, plunging Gannets that pursue quick-moving fish such as Herring and Mackerel. Guillemots take different fish to Razorbills, while Puffins sweep through shoals of small fish, taking several in a single dive. The ways in which these birds forage and feed can be helpful in identifying the species we are dealing with.

Surface feeders

One of the main styles of feeding used by seabirds worldwide is that of picking food from the surface of the sea, the bird remaining airborne or making the merest dips into the upper layer of water.

The **frigatebirds** are the only really big seabirds to practise this method. They are rangy and supremely impressive, with long wings

Seabirds have evolved many characteristic feeding techniques. Common and Little Terns (top) hover above the waves, keeping head and eyes stationary as they pinpoint a fish before diving to catch it. Manx Shearwaters cruise above the surface in search of food, but swim and dive to catch it. Shags (centre left) and Cormorants (centre right, middle) swim on the surface, looking underneath, then submerge to swim after fish, using their feet. Razorbills (bottom left) and Black Guillemots (bottom right) search underwater and chase fish in deep dives, using their wings as flippers.

and long, forked tails, combined with light bodies. This gives them a wingspan of 2.5 m or more and a body weight of a mere 1.5 kg, endowing them with great agility. They soar and sail along high over the sea, like a cross between some giant Red Kite and a Lammergeier, looking for a chance to pirate food from other seabirds but also waiting for flying fish to break the surface, or for a glimpse of a silvery fish or squid just beneath the wave tops. Then they stoop down and catch the prey in the bill, using a surprisingly long neck and a long, hooked bill to dip downwards as they pass and hook the food from the water. They perform the same manoeuvre when taking newly hatched turtles from beaches. Frigates live mainly in tropical seas, where fish are relatively few, and they may range hundreds of kilometres to find food in a single fishing trip. Nevertheless, most of their food is fish caught legitimately; piracy is the more spectacular but far less significant means of getting a meal.

Much more likely in northern waters are the storm-petrels. **Wilson's Storm-petrel** is among the most abundant birds on earth, but it is a rarity in waters where it is close enough to be seen from land in the United Kingdom. Recently, increasing numbers have been seen from ships in the Western Approaches. Wilson's Storm-petrels

Storm-petrels are tiny seabirds, but live in the middle of the oceans, pattering over the waves while taking tiny pieces of flotsam and plankton.

are small, light, broad-winged and round-tailed (although the tail often curls up at each side and gives the illusion of being forked). They flit to and fro, often lowering their legs and pattering on the surface with outstretched feet, frequently in a two-footed jump. Sometimes one will hang in the wind, wings spread and raised, feet lowered, as if standing on the sea, head bent down to search for food. In this pose they may stay still, drift backwards or do a sudden U-turn, but the broadly rounded wingtip, the tail shape and the length of leg help distinguish them from commoner species.

European Storm-petrels are much the most abundant petrels in British waters and over much of the North Atlantic. They are small, insect-like in their insignificance in the vastness of the ocean. They move quite quickly, skimming along with sideways rolls and elegant twists like swallows over a meadow. They can pace, or overtake, sizeable, fast-moving ships when they are 'going somewhere'. If they spot food, they turn sharply, or dip, sometimes using their feet alternately to patter in a running motion with wings raised. Their wings are broad-based but rather narrow-tipped and often curved back in a smooth, tapered shape; their tails are squared or slightly rounded.

Leach's Storm-petrels, a little larger than Storm-petrels, have a typically dashing, erratic flight, now shearing, now pattering or turning back on their track to capture some tiny item of food.

Leach's Storm-petrels are slightly larger birds, with longer and more pointed wings which they often hold in a slight upward kink at the wrist with the tips angled down. They glide more and even flap-and-glide like tiny shearwaters, shearing over rough seas but turning and darting around erratically. The combination of erratic movement and spells of shearwater-like glides makes this a distinctive bird at sea, and the rather long rear body, ending in a forked tail, helps to give it a different 'feel' from the dumpier European Storm-petrel. Leach's use their feet far less but sometimes demonstrate the alternate 'running' motion or even a brief pattering in some conditions.

Although far less common as a breeding bird, with only a tiny handful of colonies in the United Kingdom (compared with a wide scatter of European Storm-petrel breeding sites in the north and west), it is Leach's Storm-petrel that might be seen more often by the average birdwatcher. In late-autumn storms, Leach's, far more than European, may be seen from watching points in Liverpool Bay and along other Irish Sea coasts, and it also has a tendency to appear inland. Then a lucky birdwatcher might see the remarkable sight of a storm-petrel darting over a reservoir; or anyone may be taken by surprise and find a small seabird sitting, exhausted and bemused, in the back garden one morning after a wild night. Over a reservoir a fit Leach's Storm-petrel does not just flutter around in a small area. It may bound from one end to the other, covering a large area in a surprisingly short time, before rising up and going off rather erratically at a height, heading back to sea.

All the storm-petrels are essentially taking tiny crustaceans and plankton, as well as bits of offal and oily waste. They can be attracted to a boat by using 'chum', an obnoxious, smelly, fish-oily mixture that is often the last straw for seasick observers but brings seabirds in from several kilometres around. In such circumstances the birds feed in small flocks, and Wilson's, farther south where they are common, often gather in large groups or loose concentrations over large areas of sea. The storm-petrels compare with small song-birds on land, in having to eat and eat and eat to survive, sometimes even eating through the night. Bigger birds eat larger prey and then take a rest, but the storm-petrels just keep on finding minute items to fuel their journeys.

Shearwaters generally feed at or close to the surface, although they frequently dive (from the surface as they swim) or plunge-dive from the air to get greater depth. They move underwater with half-open wings, pursuing fish with some speed and agility, although they do not get down to any real depths. Their elongated, tapered bodies and substantial wings fit them quite well for motion underwater.

The **Manx Shearwater** is the shearwater off the coast of Britain and Ireland. It breeds in many colonies, large and small, off the

The small petrels exhibit subtle variations on a theme. Madeiran Storm-petrels (top left and bottom right) are like Leach's but with a rather less forked tail and a broader, whiter rump. Wilson's Storm-petrels, visitors from the Antarctic, are similar but even more rounder-tailed and rounder-winged. They use their longer legs in more frequent surface pattering and springy bounds.

western coasts of Cornwall, Wales, Scotland and Ireland and in some of the northern isles. Its greatest concentrations are in or near the Irish Sea, with some spectacular island sites off southern Eire, great colonies on the Dyfed islands and a particularly remarkable one farther north, on the stony, barren peaks of Rhum. This remote, lonely place may have the largest Manx Shearwater colony in the world.

At sea, the 'Manxie' speeds along in a series of flaps and glides, often 'shearing', turning over onto one wingtip, almost skimming the water surface, then tilting the other way so that it shows now black, now white. It has a fairly consistent flight pattern, with a burst of steady, elastic wingbeats, with wings often slightly arched or drooped and flexing with each stroke, then a glide that lasts as long as or longer than the active flight. In strong winds the glides get longer, the banking more marked, and the arching shears above the waves much higher and more bounding on wings that are slightly swept

Manx Shearwaters characteristically form elongated groups, or rafts, settling on the water between feeding flights, or actively diving for small fish.

back at the tip, but essentially it always has the same relaxed flaps and impressive glides. It gets along at a fair pace.

Manx Shearwaters often sit on the sea, and when they feed they will settle and dive under from the surface like auks, but never so deeply. If the fish are a little deeper, they plunge, so that they can pursue fish and squids with more speed. They are quite often found in flocks when feeding; usually, from a low viewpoint, they give the illusion of being in long lines, but elongated flocks seem to be quite common and not merely a visual effect.

In the Mediterranean, a different species, very like the Manx Shearwater and variably called **Balearic**, **Yelkouan** or **Mediterranean Shearwater**, looks a little dumpier and heavier in its actions, especially when flying low over a mirror-calm Mediterranean bay, but it is an equally elegant and fine flier in a good wind. It is a scarce bird as far north as Britain.

Rare in UK waters, the **Little Shearwater** is also tricky to identify: at least, some over-keen observers seem prone to mistake auks for the rarer bird. Perhaps it is more regular than we realize, but its true status is obscured by the difficulty of being certain about its identity far off at sea. Identification is not necessarily quite so difficult as it may appear, but getting a good enough view to offer proof, or sufficient evidence to convince a sceptical adjudication committee, is not easy.

Little Shearwaters are just that – small-scale versions of Manx, really, but with rounder wings, long in the 'arm' but short in the 'hand'. They fly fast and low, often hugging the waves and rarely rising like Manx, and they have more flap and less glide in their normal flight unless in a real gale. They flap several times with a quick, fluttering but stiff-winged action, then glide briefly, the glide often shorter than each burst of flapping; the wings stay stiff, not elastic and relaxed as in Manx. Sometimes they are described as being Common Sandpiper-like, with quick bursts then short glides on downtilted wings, but this analogy should not be taken too far; nor should comparisons with auks, as glides are much more frequent and the long, heavy body gives small-winged auks a different appearance with a much more 'whirring' effect. Little Shearwaters sometimes seem to 'bounce' off wave tops, and from ferries between the Canaries I have even watched them settle briefly with wings raised in a 'V' before simply lifting off again, stiff-winged, into the breeze.

Little Shearwaters are good divers when feeding, but they tend to be rather more solitary than the sociable Manx Shearwaters, which

The Little Shearwater is hard to identify satisfactorily as a quick 'fly-past' rarity from a UK headland, but in warmer waters, where it is more common, the dumpy form, quick flight and 'miniature Manx' appearance become easier to spot with greater familiarity.

often move about and feed in flocks (Mediterranean Shearwaters form particularly dense, active and excitable feeding groups).

Sooty Shearwaters fly north in the non-breeding season from the southern oceans, curling around the North Atlantic in a broad arc, often entering the North Sea and being seen off the east coast of Britain in late summer and autumn. They may also be seen from time to time from headlands of the western highlands of Scotland and from many Irish seawatching stations.

Sooties are rather larger than Manx Shearwaters and have a more pot-bellied look. The long, rather big body is slung below narrow, angular, almost gannet-like wings which come to a sharp point (more so than in Manx) so that even in silhouette they look different. Their flight is effortless, with long periods of glides and few, stiff wingbeats. In calm weather they may 'wander', circling and doubling back, often very low and flat to the surface, but in a gale they shear upwards and tilt over, proceeding at an impressive pace. Sooty Shearwaters are dark except for a pale underwing panel (and sometimes a discernible pale trailing edge to the wing above). They settle and dive, or plunge from a metre or so, like Manx Shearwaters, when feeding.

There are two big species of shearwater that can be seen in North Atlantic waters, but both are generally rare near British and Irish coasts. The **Great Shearwater** is, like the Sooty, a wanderer from far to the south (breeding on Tristan da Cunha in the southern Atlantic). It rarely appears in any numbers, but it may sometimes be found in the late summer and autumn in Biscay, the Western Approaches or the Irish Sea, or off the western headlands of Eire.

Great Shearwaters are very special, spectacular birds. They are strongly contrasted in pattern and are beautiful creatures, but their size and flight help put them into a different category from the smaller Manx. In active flight, Greats are like big Manx in that they flap and glide, but in a wind they shear much more dramatically, showing their larger wingspan to advantage. Their wingbeats, of course, for a bigger bird, are more powerful but less 'snappy' than those of the speedy Manx Shearwaters.

When feeding, Great Shearwaters dive more than most shearwaters, plunging in from 6–10 m in a fast, angled dive like low-diving Gannets. Usually, however, they dip in from a much lower level or settle on the surface and dive from there. They feed on fish and a good many squid.

Cory's Shearwater is the other big one, a really large, pale shearwater of the Mediterranean and a few Atlantic islands in subtropical

Cory's (upper main figure) and Great Shearwaters are super birds. Both are large, powerful shearwaters with splendid active flight, although Cory's tends to be more leisurely, even placid, in calm air.

seas. Occasionally, late-summer movements bring many to south-west England. It is a heavier bird than a Great, with a formidable, hooked bill and a wingspan in excess of 1.2 m. In fairly calm weather it gets along in a rather heavy flight, with some sideways rolling and a rhythmic rise and fall but no shearing to speak of. Its wingbeats are heavy, the wings rather bowed and broader-looking than a Great Shearwater's. In a strong wind, however, the Cory's excels. Then it rides the gale in the full glory of a shearwater in full control of the elements, rising tens of metres above the sea in immense, rolling arcs, tilting into the wind and spearing back down to the surface in a long glide before the next effortless rise. This, perhaps, is the nearest any of these northern shearwaters gets to an albatross.

Compared with the somewhat more agile Great Shearwater, the heavier Cory's is less of a diver and more of a surface-dipper and skimmer. It, too, feeds on fish and squid, but is perhaps less likely to be seen feeding in flocks than Great or Manx Shearwaters, which often gather in considerable numbers to exploit decent shoals of fish.

In comparison with the shearwaters, the **Fulmar** is a heavier, more bull-necked, broader-tailed bird. It is no bigger than the larger shearwaters but is relatively a little shorter-winged and thicker-set, altogether a bulkier bird. It is nevertheless a splendid flier.

In calm weather, it must be admitted, Fulmar flight looks like hard work. The bird progresses low to the water, flapping almost constantly in flat calm, with wings stiffly outstretched (never so languid as a large gull's). Rhythmic downbeats are broken by irregular pauses and half-glides. A little wind helps it and it glides more between groups of wingbeats, often rolling upwards and lazily tilting over, underside to the wind, turning a little around the pivot of its lower wingtip and then smoothly moving downwards again. It may turn back once more on the lower wingtip, so that it moves in a series of slow, rounded, rolling zigzags. From the bottom of the sequence it is ready to rise again to breast the breeze and continue in its rolling, tilting, slow progression.

Once there is a decent wind, it is a different story. Then the Fulmar shows off its kinship with the albatross. It moves in long, sweeping glides and broad barrelling rolls, high on one wingtip at the top of each climb before the next long, deep glide into a deep wave trough. Its wings are still held stiffly but often angled back at the wrist, looking rather narrow and pointed but without the elegant, curvy taper of a gull's wings. It also has a much thicker neck than a gull, a blunter, rounder head and a thick rear end with a broad but short tail (the lack of tail emphasized by its greyness, while the head shines yellow-white, reinforcing its size). Both in this freedom at sea and in the excitement of pairs dotted over towering, dark cliffs where they settle to breed, one can easily appreciate why the late James

Fisher found Fulmars, of all the seabirds he loved, such favourites, and why, when he entered into years of research into their spread around the British Isles, he simply became besotted with the oily, beetle-browed, black-eyed, broken-voiced, handsome birds.

Fulmars often sit on the sea, tails cocked and foreparts almost awash, or riding higher in the water with heads well up. They do most of their feeding on the water, having spotted food from the air and dropped down to deal with it more easily while afloat. They tackle quite large items and even hack big bits of carrion from dead marine mammals, using their powerful, hooked bills to rip at floating corpses of whales and dolphins or large fish. They do not, of course, have the advantage of firm ground and large, strong feet enjoyed by eagles and vultures, which can hold a corpse down while ripping it with their beaks: Fulmar prey tends to bob about, and there is no way to create a strong resistance to any upward pull.

Normally, Fulmars eat crustaceans, jellyfish and similar small fry. Fulmars have suffered from one of the more improbable kinds of pollution, as they swallow, in mistake for jellyfish, many of the innumerable rubber contraceptives that are washed out to sea from sewers; they can be lethal.

If they find a shoal of fish, Fulmars are quite capable of diving from the surface like shearwaters, using their wings to propel themselves several metres deep. Plankton-rich areas, offal around whalers and trawlers and other temporary honeypots attract hundreds of Fulmars, often in the company of gulls, skuas and Gannets, sometimes in a blinding whirl of spray and snow or freezing rain. They are opportunistic and so have thrived in the modern world where man's waste at sea includes much that is edible to a seabird with a catholic diet.

It is every seabird-watcher's ambition (at least, it is mine, and I do not think that I am too unusual) to see an albatross. In the coastal waters of Briain and Ireland, the chances are that if an albatross is seen it will be a **Black-browed Albatross**. One has been sitting in a gannetry on Shetland each summer for a number of years and occasionally it (we presume) is seen elsewhere (usually at sea, from a ship or an oil rig).

Albatrosses are very large to huge; a Black-browed is at the very large (that is to say, small) end, not such a giant as a Wandering or Royal Albatross, but is bigger than a Gannet, nevertheless. It is also a different shape from a Gannet, more like a giant, long-winged Fulmar with a longer, thick bill, a large head and humpback, and a short, broad, squared tail. The wings are extremely long, with the inner wing noticeably comprising upper arm and forearm (more so than any Fulmar or shearwater), giving a stretched 'Z'-shape rather than a 'V'-shape to the angled wing. It flies, in a wind, with long, shearing glides and hardly a wingbeat at all. In calm weather, like other big albatrosses, it prefers to sit it out on the sea, where it looks big, rides high and has its tail well clear of the water, like some enormous, short-necked goose.

Plunge-divers

Once food is more than a few centimetres below the surface, a bird has to dive to reach it. Already we have seen that some of the 'surface feeders' are capable of diving and that the surface dippers intergrade with the shallow-plunge-divers.

Plunge-diving is diving in from the air, rather than from the surface: a dive from the high board, not simply sinking under while swimming. Some species dive from a great height and go in with a terrific splash, having special adaptations to prevent injury from the severe impact. Others are lighter birds that dive from lesser heights, but they still strike the water with a loud smack and a jet of spray.

Some other birds dive, but they are not plunge-divers in the sense that these seabirds are. The Osprey, for instance, dives for fish and often totally submerges, but it goes in feet-first, using its sharp claws and muscular legs and feet to grasp and subdue its catch. Closer to the seabirds in technique are the kingfishers, many of which hover over water before diving in head-first, while others usually dive from a perch.

The seabirds dive in headlong and use their bills to catch their prey. To work well, the plunge-diver has to have a bit of weight behind it and a streamlined shape to help it penetrate the dense seawater. In the USA, the Brown Pelican can be seen diving in spectacular fashion, but it is a little unwieldy for much more than superficial, shallow diving. It uses a Gannet-like dive (it is, in fact, closely related), using wings and tail to control its angle of dive and to aim at the fish, which it watches all the way down during its descent. At the last moment, however, it pulls its long neck backwards and its head is drawn back above the shoulders, so it never has the streamlined look of a Gannet. What it does have, though, is an enormous, elastic pouch which expands as it hits the water and surrounds the fish, while the upper mandible snaps shut to prevent escape.

Tropicbirds plunge-dive, but they do not have the size and weight to take them far under. It is the speed of the dive that propels them, as they use neither feet nor wings while submerged, and the smaller species just do not generate sufficient impetus to get very far.

The technique is shown to perfection by the gannets and terns, each in its different way, but many seabirds and other species have a go from time to time. Even a Grey Heron might drop heavily into the centre of a reservoir to try to grab an unwary fish, while Shags sometimes drop out of the air and dive under. Gulls are rather more accomplished, doing most things quite well; they plunge often, but they are too buoyant to submerge very far.

Gannets are exceptional birds. In British and Irish waters they are the biggest of the regular seabirds, 1 m long and approaching 2 m across at full stretch. Adults gleam white, set off by big black wing-tips and a shawl of bright golden-buff, the coloured neck feathers looking soft and velvety. They are at home in the roughest conditions, rarely being blown far off course or inland: they are powerful and confident creatures of the open ocean.

In comparison with other gannets and the similar boobies, the **Northern** (or Atlantic) **Gannet**, is the biggest by a considerable margin. This gives it added weight, which means more dramatic and effective plunge-dives. It is long- and narrow-winged, and the wing shape is such that the upper arm and elbow are much more fully visible than on most birds, giving three segments of wing rather than two, much more like the shape of a human arm (counting the hand).

Gannets soaring beside their nesting cliffs show magnificent precision and delicacy of adjustment in their wing shapes and angles. Circling out in calm air, the wings may be at full stretch, the leading edge dead straight, the trailing edge curving forward to the tip. Then, heading to the cliff and gaining speed, the bird will flex its wings, the tips angled slightly back and much more tapered; a banking pass close to the rocks calls for air to be slipped between separated primaries and from beneath the uptilted tail, allowing slow-speed manoeuvres without a stall. The tail is fanned and closed, and the legs tucked away or drooped, bent or straightened, toes spread wide or tightly closed, to increase or reduce their surface areas. Coming in to land, a Gannet uses all the available area of angled wings, fanned tail and raised feet to slow it down before impact and all its accumulated experienced and practised tricks to get it safely down at the desired spot.

When actively looking for fish, Gannets fly at height (10–30 m depending upon weather and surface conditions), heads tilted down as they scan the waves below. The flight is powerful, not fast, with regular, strong downbeats of the long, pointed wings. The wings may be straight or somewhat angled, the tips flexing upwards on the downbeat, adding to the impression of size and strength. In a wind, flight is broken by glides; downwind it flies direct, gliding level and straight between groups of three or four or half a dozen wingbeats. Across a wind, or when flying lower over big waves, it looks like a large shearwater, banking onto its wingtips, shearing upwards breast to wind and then sideslipping down again, with infrequent beats or none at all for minutes on end. It is a master of its element.

Feeding involves the intent downward scanning, then a sudden rise, even a half-turn, to come to a halt above the intended prey. If prey is

A deep fish requires a full-tilt, headlong plunge that enables the Gannet to reach a sufficient depth to catch its prey. At the last moment, it swings its whole wings backwards, like a streamlined arrowhead.

spotted farther ahead, the Gannet confidently flies downwards in a fast power-dive, or simply pulls in its wings and swings into a long, slanting plunge. Many dives are from a low elevation, and by no means all are the spectacular, full-height, textbook performances.

The dive is controlled by flexed wings and the pointed tail, allowing fine adjustments on the way down while the bird keeps its fish in view. The feet may also help as rudders or brake the bird before impact. To the naked eye, at a distance, the bird seems simply to close its wings and spear into the sea with a great splash. In fact, as can be seen well at closer range, and even better on film, at the last second it pulls back the innermost part of its wings – equivalent to the upper arm – tight along the body, the forearm (the 'inner wing' on most birds) angled out and the tips bent back, like an arrowhead.

This is real power and speed combining with the weight of the big bird to take it as much as 10 m deep. Often the depth is much less and the Gannet pops back up within a matter of seconds – rarely more than ten seconds elapse. The fish may be caught in the bill on the downward route, or snatched from below on the return to the surface: it is usually swallowed before the Gannet gets back into fresh air.

If fish are abundant close to the surface, Gannets may plunge in shallower dives, knifing in from a metre or two at a low angle. Sometimes flocks rain down on big shoals in non-stop splashing, crashing descents, one of the great seabird spectacles in the North Atlantic.

Compared with Gannets, terns are lightweights, almost amateurs in the art of power-diving from a height. Yet a Sandwich Tern puts on a good show, and others, even if only shallow divers, are experts in judging a dive to perfection to give the desired result: a captured fish.

Terns in general are slender birds, more so than gulls, with narrower wings and usually more elongated and deeply forked tails. Their flight is buoyant and elegant, to a greater or lesser degree, but, oddly perhaps for such aerial birds, they virtually never glide except when coming down to land or in special display flights. A gull following a ship will glide continuously; a tern will beat its wings continuously.

Sandwich Terns are exceptionally pale, standing out in most light conditions at long range. They are rather large, with a long, thickish-based bill, frequently held level in direct flight but angled down when hunting. The head is roughly crested at the nape, but this is rarely noticeable in flight, when the feathers are smoothed down. The body is not over-long for a large tern, without the smooth, deep-chested bulk of a Caspian or Royal; in fact, it can look a little stubby at the rear, an effect accentuated by the relatively short (but deeply forked) tail. The wings are long and narrow, usually showing a marked bend at the wrist, and they are set well back on the body.

Normal flight is strong and direct, with regular wingbeats that are not particularly deep and are somewhat stiffer than in smaller

Depending on the depth and speed of the fish, Gannets use a shallow dive at a low angle, speeding into the water to scoop up its prey.

species. In strong winds the Sandwich Tern looks quite comfortable, often slipping sideways and down to the wave tops before careening up again. Only juveniles, with shorter, rounder wings and tails, look inhibited and lacking the power to cut through the turbulent air.

A feeding Sandwich Tern flies at a height of perhaps 10 m above the sea, often higher than Common Terns, looking down as it flies. On spotting a likely fish, it pauses, perhaps rising a little to take off the forward speed and position itself in a brief hover above its prey. Then, quickly, it is into a steep, often vertical, dive, hitting the water head first with a loud, abrupt smack that sends up a noticeable jet of water. The hover preceding the dive is short, and sometimes dispensed with altogether, rarely so consistent as in Common Terns. In a high dive, it may use a wingbeat or two to adjust its angle and position. It usually just submerges, but often the wingtips, at least, remain above water; the fish is brought to the surface held crosswise in the bill.

Sandwich Terns rarely swoop and dip in the manner of smaller species, although they can, and they may even catch insects over land on occasion. Sometimes they settle on the water, swimming buoyantly but usually briefly before moving on. If a flock on a beach is disturbed, perhaps by a falcon or skua, they will fly off, with jinking, dipping

twists and rapid, flickering wingbeats that take them away at speed and confuse the eye of the attacker in a dazzle of silver and white.

Much rarer, indeed now one of the rarest of all Atlantic seabirds, the **Roseate Tern** is equally pale and striking at sea. It is closer to a Common Tern in many respects, yet has the look of a Sandwich Tern in its long, spiky bill, silvery upperparts and white underside. Both these species make even a Common Tern look grey and dingy.

Roseate Terns were misrepresented in field guides for years, being accorded a particularly graceful, elegant flight. They are, without doubt, graceful birds of a special beauty, but their flight is not so elegant as that of many of their relatives. The Roseate Tern is long-bodied, long-billed and long-tailed, but the wings are rather short and stiff. It appears sometimes to have a quite broad, rounded head, perhaps an effect of the extra contrast of its jet-black cap with whiter underparts than on a Common or Arctic Tern. The long tail streamers may be held together in one lengthy, tropicbird-like spike.

Its flight is closer to that of a Little Tern than that of a Common or Arctic: not quite so frenetic, but similarly quick-winged and stiff. Its wingbeats are shallow, without the lovely flexibility of an Arctic Tern's. They appear stiff and show equal emphasis on both upstroke and downstroke, subtly different from Common and Arctic Terns.

Roseate Terns cover long distances between dives, and are well spaced even when many are in an area. They fly higher than Commons or Arctics when fishing, often between 8 and 15 m. If they join a dense, mixed feeding flock, they are often at the edge and low down.

When looking for prey, Roseates hover briefly, assisted by a good wind but otherwise seem to struggle to keep in place for more than a second or two, with rapid, almost whirring wingbeats. On sighting a fish, the Roseate Tern simply changes its angle of flight and flies downhill, hitting the water 'on the run', diving at a greater speed than other similar-sized terns. This allows it to penetrate a little deeper and remain submerged for longer, often two seconds or more.

Common Terns are generally common in inshore waters from spring to late summer and frequent on through the autumn. **Arctic Terns** are more common farther north, where they can sometimes be abundant, but in some southern parts of Britain and Ireland they are quite scarce migrants. Both are greyer birds than Roseates. Common Terns have rather long, stout bills, a relatively long head shape (unlike the round-headed Arctic) and almost as much in front of the wing as behind. The Arctic has a neckless appearance and a longer, more attenuated rear end, so it seems to have more behind than in front.

Commons have fractionally broader wings than Arctics. The inner wing (arm) is perceptibly longer, the outer part subtly shorter and blunter. The Arctic's wing is particularly short-armed, and the long tip is tweaked out to a narrow, sometimes slightly backward-curving point, adding another extra touch of refinement. Together with the head-and-bill length, this gives the Common a somewhat different, more cruciform shape and a very slightly heavier, less wonderfully elegant look than an Arctic Tern. The superlatives have to be added to the Arctic, because to differentiate it from the Common by saying that the latter is 'heavier' is to stretch the point somewhat: the Common, too, is a light, airy, delicate bird. It has a fast, strong downstroke and a slower, less emphatic upstroke, whereas an Arctic has an upward 'snap' followed by a slower downbeat. This sounds, and is, subtle; but, if you watch a flying tern and try to see these differences, it is easy to see the downstroke on an Arctic and less easy to observe the quick upstroke, whereas it is slightly the reverse on a Common. They are tricky, these two, but it is always good fun trying to sort them out.

Both Commons and Arctics actually get along more slowly over the sea than Roseates, which fairly whiz by. Both will gather in feeding flocks or even spiralling funnel-shapes centred above shoals of fish. They fly at 3–8 m (lower than Roseates), heads bent down to spy their prey. If a fish is seen, a Common Tern will turn back and upwards to reduce forward motion, then hover or simply dive straight in. An Arctic Tern tends to perform a characteristic kind of dive, peering down from a hover, dropping on still wings, then pausing again to hover lower down, as if checking the position of its prey. It may dive, or hover at yet another level, or simply peel off and move away.

This less confident, hesitant performance is usually quite distinct from the fast plunge of a Common Tern or the power-dive of a Roseate. Commons and Arctics usually work a small area thoroughly, with many dives in succession, unlike wide-ranging Roseates.

Both Common and Arctic terns frequently also dip to the surface to pick insects, crustaceans or floating fish fry from the waves, and they will also catch insects in the air. In these activities they appear much the same in their actions, although the Arctic generally retains its especially light and buoyant appearance.

In North America (rarely in Europe) are **Forster's Terns**. They resemble Common Terns, but more often capture insects in the air and have a lively, buoyant look, with shallow wingbeats (although these can look markedly angled at the wrist, giving a flattened 'M'-shape).

Much smaller than an Arctic Tern, the **Little Tern** is the most frenzied of all the terns. It is quite bulky-bodied and thick-necked, with a spike-like bill, narrow wings and a fairly short, narrow tail. Unlike Common, Arctic and Roseate, its tail lacks streamers at each corner.

A Little Tern on the move usually flies straight and true, with constant, quick wingbeats. Its flight is light and airy but without the fluency of the larger terns. It may dip to the sea or the edge of the beach as it goes, but it is clearly a bird that knows where it is going. When fishing, though, it may turn and twist and look again and again at a suitable patch of water. It hovers with whirring, but deep, wingbeats and frequently drops abruptly, to hover a bit more, or it may abort the dive and move away. It often dives at the very edge of the surf off a steeply shelving shingle beach. Like the Sandwich Tern on a smaller scale, it goes in hard and fast with a sharp splash.

Caspian Terns are the biggest. They are larger than Common Gulls, but smaller than Lesser Black-backeds or Herrings and not so heavy, although they look long-winged and are often described as 'Herring Gull-sized'. Their flight is more gull-like than that of smaller terns, with strong, slow wingbeats and frequent glides, and they look more bull-necked, but do share the vertical plunge for fish. The dive is often high, fast and dramatic. They are generally scarce along western European coasts, and in British and Irish waters always very rare.

Almost as big and impressive, but rather longer-bodied, longer-tailed and more slender-billed, **Royal Terns** are rare everywhere in the North Atlantic, being more subtropical and tropical in distribution. They have long, rather narrow, angled wings; Caspians are broader-winged, although from some angles they can look very Gannet-like in shape. All the same, Royals tend to appear like giant Sandwich Terns rather than smaller Caspians, more tern-like than gull-like overall.

Even rarer, but with a recent tendency for one or two to reappear in the north, **Lesser Crested Terns** are like somewhat smaller Royals, with wings about as long as a Sandwich Tern's but proportionately

A fishing Common Tern has an air of determination and confidence. It hovers, moves on, hovers again, then, spotting a fish, goes down in a clean, direct dive. Usually the result is a small, shiny fish held in the bill as the bird emerges; it is quickly swallowed or taken back to the colony.

Arctic Terns look less assertive than Commons when fishing. The dive is often preceded by a hover, a dip, a further hover, a sideslip or another small adjustment before the final quick, smooth entry. This hesitant, 'stepped' dive is a useful identification feature even at very long range.

broader, giving a more even, flowing flight action without quite the same bounce and springy wingbeats of Sandwich Terns.

Gull-billed Terns are coastal and marshland terns which feed over wetlands or damp meadows rather than the sea. They are often seen over the sea close inshore and are distinguished from the Sandwich Tern by their marginally broader, blunter wings, although this can be overemphasized, and they are frequently perceived as long-winged with tapered, pointed wingtips slightly swept back.

Normally, Sandwich Terns plunge, but Gull-billeds are not plunge-divers, but surface feeders in active, dipping flight. They are graceful, swooping, terns, but lack the hover, plunge and splash of real sea terns.

Barely meriting a mention here, **Sooty Terns** are exceedingly rare in the North Atlantic. They are large, long-winged, full-breasted terns with particularly long tails and a strong, fluid flight. They dip to the surface to feed and rarely swim, staying aloft for months, even years,

A Lesser Crested Tern looks like a broad-winged, dark, Sandwich Tern with a bright orange bill, or a rather small, slender Royal. A single bird out of range is difficult without a good size comparison with another species alongside.

The Gull-billed Tern (left) is somewhat bull-necked and stubby-billed for a tern, but still has long, tapered wingtips. The Royal Tern is bigger, long-winged and rangy, marked out by its large orange bill.

at a time. **Bridled Terns** are equally rare, rather smaller and less long-winged and long-tailed, but their flight is especially elegant and elastic.

Other terns are seen at sea, but are normally thought of as freshwater-wetland birds: the **Black Tern** and **Whiskered Tern**. Both are unlike the sea terns in shape and action (although the Whiskered, especially, is often claimed from a poor view of a young Arctic Tern). Black Terns are regular coastal migrants. They are small, lightly built birds without the long tails of the sea terns. They never adopt a hover-and-dive fishing technique, preferring instead to fly steadily, low over the waves, dipping to the surface to pick up floating food. They sometimes rise higher and circle in a slow, cruising soar.

Gulls

Terns are the elegant seabirds, the ones that everyone likes even if they make screeching noises and dive at intruders' heads. Gulls are the good-for-nothings of the bird world, often ignored, disliked even, and frequently described as either 'boring' or 'too difficult to bother with'. At least when they are 'a noisy nuisance' they are taken notice of. In fact, they include some stunningly beautiful birds and are fascinating, instructive and always entertaining to watch.

Some seabirds are specialists. A Wilson's Storm-petrel dances across the waves; a Gannet plunges from a great height; a Little Tern slaps into the breaking waves at the edge of a beach. None of them can do much else. If their food supply dries up, they are stumped: you will never see a Gannet foraging on a beach or a sea tern following a plough.

Gulls are different. Most of them can turn their hand to almost anything. Sabine's Gulls, Audouin's Gulls and one or two others are pretty specialized, but the rest may be seen in a variety of situations, responding to opportunities, both natural and artificial. Herring Gulls follow ships far out of sight of land; they feed at sewage outflows; they turn over the seaweed at the edge of a beach; they feed on tips far inland and even come to pick up scraps on popular mountain tops. They, and several others of their kind, are remarkable and must be admired for their ability to succeed in a changing world.

Gulls vary in size, from one or two smaller than a Common Tern to birds as big as more impressive birds of prey, dramatic, powerful birds that fly like bombers and swim like battleships. Great Black-backed Gulls are huge in comparison with, say, Black-headeds and Commons; even next to a Lesser Black-backed Gull, not many inches shorter, they can look 'twice as big' and they weigh twice as much.

The variety in the gull group is considerable, and this is reflected in their shapes, flight actions and ways of life. Yet all look like gulls: something about them, and particularly the predominance of white in the adult plumage, makes them instantly distinctive, even if it may be tricky to sort out precisely *which* gull it is.

The **Little Gull** is the smallest. It breeds inland in northern Europe but winters at sea, with large concentrations in the Irish Sea, birds which pass by the north-west English coast in spring. Little Gulls are surface feeders over water. They have a quite tern-like flight, but, even though their progress is varied by dips and sideways turns, they have not quite the character of a sea tern – more, perhaps, that of a Black or Whiskered Tern over a marsh.

Adult Little Gulls have quite long but blunt-tipped wings; the tips are much more pointed in individuals under a year old. Young birds have slightly forked tails, although this is hard to see, but adults have short, neat, square or rounded tails. All have particularly small heads and slender bills, giving a gentle facial appearance.

The flight is low, slow, dipping, with quite quick wingbeats that become rather jerky and irregular in a wind. They frequently beat into wind for a long way and then turn and run back with it before retracing their steps, dipping to the surface to pick up tiny items as they go. Often a small group will feed like this together, although not well coordinated in their actions. Groups of birds also sometimes circle together and gain height in a spiralling flight which may end in a high-level movement away or a slow descent, the flock breaking up.

Black-headed Gulls are familiar in the eastern Atlantic but still baffle some people with their variety of plumages. The adult in summer (bottom left) is easy but a one-year-old bird (bottom right) is less obvious. An adult Little Gull (middle) in summer shows a distinctive dark underwing; Bonaparte's Gull (top) is a rare visitor from North America to Europe.

Almost as small, but more slender, the **Bonaparte's Gull** is a New World bird, rare east of the Atlantic. It is much smaller than a Black-headed Gull, but on a lone bird this is difficult to judge. It also has a slim bill, quite long and pointed, even thinner than a Black-headed's. In flight it is long-winged, the wings often angled and drawn back to a sharp point. In steady progress into a wind it recalls a Little Gull, or a tern, being light and lacking the power of bigger gulls, but it is agile and capable of quick and complex aerial chases for insects.

Black-headed Gulls are abundant east of the Atlantic and around the North Sea and (in summer) the Baltic. They are light, small gulls with tapered, angled, pointed wings and insubstantial bills. Normal flight is direct, with regular, quite shallow wingbeats, but it can be varied according to conditions and intentions. In a gale, Black-headeds fly head-to-wind with confidence, riding the currents beautifully and rarely looking at all discomfited by the wind. When feeding from the sea or taking food from newly turned earth, or newly tipped refuse, the active flight becomes quick and agile, with fast wingbeats, hovers, dips and twists. They can also rise high in the air to catch flying ants, but they do not often soar or glide great distances.

A little stockier in build than a Black-headed Gull, and typically only a fraction larger, the **Mediterranean Gull** is a very much scarcer bird. It is, however, regular in small numbers around the southern North Sea and the English Channel, with a few individuals farther west. Compared with a Black-headed Gull, a Mediterranean has a bigger, chunkier head (broader-crowned and perhaps a little squarer in outline) with a thicker, although often not longer, bill. The tip of the bill is thick and sometimes has a drooped look, or a more blob-shaped effect than the tapered bill of the Black-headed Gull.

Its wings differ, too, being a little straighter, broader and blunter, especially in adults. It holds its tail closed in direct flight, but when feeding it is equally as adept at aerobatics and low-level plunges as a Black-headed and in such activities the precise shapes of wings and tail are constantly changing. The differences are slight and picking them out takes a practised eye, but an experienced observer can single out a Mediterranean Gull from a Black-headed at long range, even at a glance. Even on the beach, the confident swagger, big chest, flat back and long legs of a Mediterranean distinguish it quite well.

There are two rare vagrant gulls from North America that look a little like Mediterranean Gulls in summer, when adults have similarly jet-black heads with white eyelids. One is a touch bigger and more slender: the **Laughing Gull**. It has a long, slim bill that may appear to droop (closer views can dispel this impression), and long legs which give it a free, high-stepping, quick action on land. In flight it is particularly rakish, its long wings tapered to sharp points.

Franklin's Gull is smaller, somewhat dumpier, and blunter in the wing. It has a shorter, relatively thick bill with a tendency to a more abrupt downturn at the tip, and shorter legs making it less active ashore. In flight, the quite broad and square-tipped wings have a paddle-shape and a quick, stiff action that can be characteristic.

Common Gulls are bigger than Black-headeds but smaller than Herring Gulls. They are intermediate in many ways. The bill is thicker than a Black-headed's but, in comparison with the heavy hook of a Herring, slender and weak. The whole head-shape is genteel, with a steep forehead and round crown above a big, doey eye. On the ground, the Common Gull looks long and tapered: its wingtips extend well beyond the tail (especially on young birds), giving a more elegant shape than the Herring Gull. In flight it has long, tapered, angled wings, but they come (on adults) to a slightly blunted tip. This is in part due to the pattern, but is nevertheless a feature of constant value.

Common Gulls are really fluent, beautiful gulls in the air, but have neither the power nor the gliding ability of Herring Gulls. They are neither so quick and nervous as Black-headeds nor so assertive and self-confident as Herrings, moving with a lovely action lacking excesses or special attributes. On the ground they are nimble, but shorter-legged than their American equivalent, the Ring-billed Gull, or the Mediterranean Gull, and not so long-striding and aggressive.

In North America **Ring-billed Gulls** have increased dramatically in recent years along Atlantic coasts, and this has been reflected in a greater frequency and consistency of occurrence in western Europe. Ring-billed Gulls are clearly close to Common Gulls but a touch on the Herring Gull side in terms of size and shape. The thicker bill looks as if it is slightly open, holding something along its length, compared with a Common's, being just that bit stouter and blunter-tipped. The head can be flatter or squarer, although it can equally well look perfectly rounded: describing gulls' shapes is fraught with difficulties of semantics and perceptions and, in any case, they change with the birds' moods!

Like Mediterraneans next to Black-headeds, Ring-billed Gulls compared with Commons are broader-winged, longer-legged, bolder-chested and more upstanding. On the ground they are long-stepping, quick-marching birds, confident and strong. In the air they are strong fliers, with broader wings tapered more to a point than Common Gulls' wings, giving a distinctively different silhouette; consequently, they have a fractionally stiffer action and a more forced look about them.

Herring Gulls are bigger than Ring-billed or Common Gulls by quite a way: they are noticeably 'big' rather than 'medium' gulls. As with other gulls, the male is the larger-headed of the pair, but this becomes more obvious with the larger species.

Herring Gulls are stocky, heavy, long-winged gulls; compared with Lesser Black-backed Gulls, they are somewhat shorter-winged and longer-legged. The bill is stout and noticeably hooked. On the ground they stand horizontally or with a more sloping back, but both Lesser Black-backs and Common Gulls are, on average, with their shorter legs and longer wingtips, more sloping down to the rear.

When feeding ashore, Herring Gulls are aggressive and dominate smaller species, often flying up, half hovering on raised, arched wings and plunging down head-first to dive on to food. They will also dive on to competitors, pecking them viciously about the head and body.

In flight, Herring Gulls are masters of their element. They call a lot in flight and frequently stretch the head and neck far forward as they do so, creating a peculiar silhouette. Along a clifftop they glide to and fro all day long. Behind a ship they slipstream with consummate grace and effortless ease. In a gale they ride the wind and shear over waves like shearwaters, and in calm weather they may circle and rise high on thermals like big birds of prey. They are also nimble and accurate, despite their heavy, powerful nature: anyone who has lost half a sandwich to a Herring Gull can testify to that.

Lesser Black-backed Gulls are almost as big and weighty as Herring Gulls, but on average a touch more elegant and long in the wing. They are equally good at soaring, sailing, gliding, diving and powering after other gulls to steal food in mid-air and are, if anything, even more predatory when it comes to attacking and killing smaller seabirds and catching their own fish. They plunge-dive from low level, hover, twist and turn in no time at all.

Like Herring Gulls, Lesser Black-backs feed at tips and loaf about on nearby wasteland, playing fields or beaches. They roost on inland waters, often flying to them at dusk in long lines and 'V's, almost like skeins of geese. If disturbed at roost, they rise up in swirling flocks against the sunset, creating some of the most dramatic birdwatching spectacles in inland Britain; at their coastal colonies they are scarcely less impressive in flocks, and always beautiful to watch individually.

The most frequent American gull on British beaches is the Ring-billed Gull (left), which has a somewhat intermediate appearance between Common (centre) and Herring (right) Gulls: note the leg colours, eye colours and bill patterns and the Common's darker back.

Scandinavian Lesser Black-backed Gulls are even longer-winged and more slender overall, rather smaller-headed and more finely balanced. These are long-distance migrants that have evolved longer wings to help them on their mammoth journeys, perceptibly different from the British birds that travel less far on their migrations.

In the Mediterranean and increasingly on English Channel and North Sea coasts, **Yellow-legged Gulls**, or Yellow-legged Herring Gulls, are found. They are confusingly intermediate in many respects between Lesser Black-backs and Herring Gulls, but several races have different characteristics that may turn up and generalizations are dangerous. Mediterranean Yellow-legged Gulls are big and heavy; others are long-winged, round-headed, smaller-billed and elegant.

While Herring Gulls may dwarf Common Gulls, they are often not much bigger, if at all, than Lesser Black-backs (they are all bewilderingly variable). **Great Black-backed Gulls**, on the other hand, certainly look bigger than Lessers, and some big males beside small female Lessers simply look like giants by comparison.

Great Black-backed Gulls are heavily built, massively muscled, with big heads and huge, deep, heavily hooked bills that make short work of a Puffin or a Manx Shearwater. Even obscurely patterned young birds have the distinctive, thick lump of a bill that is never matched by a Lesser's. They are also broad-winged and wide-bodied.

This is often seen to advantage when a mixed group of gulls is watched in a field beside a refuse tip: a Great Black-back will fly in, gliding down to settle by its neighbours, looking likely to squash them flat as it lands.

In flight over the sea, Great Black-backs are slow and lumbering with ponderous wingbeats, but they are also capable of fast dives in a mêlée around a trawler or above a sewage pipe, and are aggressively dominant over the other species. They take off from sea or land with a characteristically heavy, strenuous action as if barely able to get into the air, but once airborne they are very able fliers. Like Herring Gulls, they soar above and glide alongside coastal cliffs, riding the upcurrents to great effect, using no energy in their territorial patrols.

At coastal sites and inland tips in winter, two 'white-winged' species are found. The **Glaucous Gull** is the equivalent of a Great

In Iberia and along Mediterranean shores, but increasingly also farther north, Yellow-legged Gulls can be found. They have usually been considered to be the same species as the pink-legged Herring Gull, but both breed in some colonies without hybridization.

A Great Black-backed Gull in its first year (top left) is chequered brown and white and it is four years before it becomes neat black and white as an adult (right). Glaucous (bottom left) and Iceland (bottom right) Gulls look alike, always with pale wingtips and tail; in this first-winter plumage the larger, pink and black bill of the former is a good distinction.

Black-backed, the Iceland Gull more the equivalent of a Lesser Black-backed, although Glaucous are often only 'large Herring Gull'-sized. Glaucous Gulls are Herring or Great Black-back shaped, being big, lumbering, broad-winged and large-headed. Nevertheless, they look fine, soaring well, or flapping low over the waves heavily but elegantly.

Iceland Gulls are essentially smaller versions of Glaucous, with near-identical plumage sequences (from barred brown juveniles to pale grey and white adults). They have shorter bills and a gentler facial appearance, but too simplistic an approach is fraught with danger. True, Glaucous may look mean, flat-crowned, large-headed and long-billed, but they can also be round-headed, too. Icelands have steeper foreheads, less 'chin', higher, domed crowns and shorter (often stout) bills, and longer wings that look more tapered beyond the tail at rest. There is often, but by no means invariably, a 'step' effect at the rear of the back (the tertials) above the bunched wingtips, which project at a lower level, whereas on a Glaucous this is more often a smoother profile, or a bigger, long, rounded bulge or hump.

Although smaller, Icelands are short-legged and podgy-bodied, with round chests and pot-bellies, reflecting the thickness of plumage appropriate to an Arctic species. Some juveniles look very long-winged when they swim, the wingtips tapered and cocked upwards in an extravagant manner never suggested by a Glaucous, while adults can be shorter-winged, especially early in the winter, and not so obvious.

In flight Icelands have broad wing bases and heavy bodies (some early field-guide descriptions comparing them with Black-headeds in terms of lightness of flight and narrowness of wing being misleading), but the wingtips may look longer and narrower than on a Glaucous.

When feeding, Iceland Gulls are usually more aloof, less likely to leap into the wheeling, squealing scrum of birds fighting for food, more often noticeably lower in the hierarchy than the big Herring Gulls, Great Black-backs and Glaucous. All the same, these are tricky birds, not easy to separate on brief acquaintance.

At a coastal-cliff colony, one gull often dominates the scene: the **Kittiwake**. It is a glorious bird, with row after row of squealing gulls on minute nests clinging to the tiniest of ledges.

Kittiwakes are much more ocean-going than most gulls, more truly seabirds in the way that shearwaters and Fulmars are. Many spend the winter around harbours (at least they do not visit tips) but most are out at sea, where a seagull should be.

Kittiwakes are very short-legged and long-winged, so they are poor walkers on land; they do sit on flat beaches, looking heavy and tail-down, almost pear-shaped, but more often they perch on a cliff ledge in a rather upright fashion. In flight, they show long wings of rather uniform width all the way down to a slightly blunted tip (the leading edge particularly straight), although once flexed back a little (as in fast flight over the sea) the wingtips look very pointed. A distinctive feature, especially at the breeding cliff, is the effect of 'as much in front of the wing as behind', with a rather long head-and-neck profile, a short body and a short, narrow, square (even notched) tail.

Juveniles have less of this effect, in fact looking hunched with short necks and heads tucked back into the shoulders. They have broad-based, tapered wings giving a more triangular impression. At sea, Kittiwakes are consummate fliers, sailing beautifully in a breeze and flying fast in a gale, with a dynamic, bounding rise and fall and shearing on one wingtip like the Fulmar or Manx Shearwater.

Sabine's Gulls are similar, equally extrovert, but they have slightly broader-based wings leading into long, tapered wing points and more deeply notched tails. Kittiwakes plunge and dip into the waves; Sabine's Gulls dip and also feed from the surface as they swim, with heads well up, bills turned down, tails cocked. If there is a strong wind, a Sabine's might head into it, wings wide-spread and flat, and patter over the waves with its feet as it feeds, almost like a storm-petrel.

In direct flight Sabine's Gull has a tern-like appearance, with the long wingtips emphasizing the similarity. The wingbeats are deep and rhythmic, giving a wavering flight low over the waves on migration. In strong winds it seems to seek shelter behind big waves, twisting and dropping into deep troughs, stalling, or sweeping up on to one wingtip and letting the wind take it across into the next trough.

Kittiwakes are common; Sabine's are rare and frequently the object of wishful thinking at the sight of a distant Kittiwake in the autumn.

Ivory Gulls are rare, exciting, medium-sized gulls with round heads, rounded, deep-chested bodies and long, paddle-shaped wings. They are birds of the pack-ice edge and rarely stray south. Their flight is beautiful, with elastic wingbeats and marked aerial agility, but also periods of stiff-winged soaring and gliding. In a gale over the ocean, the Ivory Gull is capable of steep, fast careening flight like the Kittiwake.

Ross's Gulls are also Arctic gulls, although they nest south of the tundra, and are very rare and exciting finds, chiefly in the North Sea. They have lovely round heads with tiny bills and are short-legged and almost dove-like on the ground. In flight, they look long-winged, but the wings are fairly broad-based and tapered to a point, often angled back. The tail is uniquely wedge-shaped, but this is not easy to see. Their flight action resembles that of a Little Gull or Black Tern, with shallow wingbeats, hovers and sudden dives or dips to the surface. In a wind, if the bird is moving longer distances, a shearwater-like flight can be adopted, or it may move with quicker, deeper wingbeats.

In the Mediterranean and on the west coasts of Iberia, but no farther north, **Audouin's Gulls**, now afforded greater protection, are increasing after years of being among the world's rarest seabirds. They look superficially like Herring Gulls, but their distinctive head shape (with a long 'snout' effect) and bill (thick, parallel-edged and blunt-tipped) help to distinguish them at close range. At long range the 'blob' bill may still be apparent, but the long, narrow, angled, pointed wings (almost Gannet-shaped) and short, narrow tail may then be the best structural clues. On land, they look impressive, self-assured, upright gulls with heads held high, back sloping and wingtips and tail tilted downwards in the manner of a Ring-billed Gull. Unlike Herring Gulls, which feed on anything, anywhere, Audouin's are more specialized, usually catching fish from the surface of the sea as they fly, but recently some seem to have taken to foraging on beaches and even tips.

Deep divers

From the point of view of a birdwatcher on shore, or even aboard ship, it is the plunge-divers and surface feeders that appear most interesting and dramatic: simply because they are so visible. The birds that swim on the surface and duck under to hunt have less immediacy, as their behaviour underwater is hidden to us. Only in good conditions, from the top of a tall cliff above clear water, can we sometimes see auks performing underwater for a short way, to get just a little idea of what life is like for them out at sea.

These divers, however, include some remarkable performers. Compared with the plungers, they go much deeper, using their wings or feet to propel themselves through the water at great speed and with real agility. They are never fully matched by the superficial efforts of the plungers, which simply go in as far as their airspeed takes them and then inevitably bounce back to the surface like corks.

The problem of underwater pursuit is the power needed to force a way through dense seawater. It is simple for us to run along a beach; not so if we are up to our knees in water, and worse still if we go deeper and have to counter the tendency to float as well as raise the effort needed to move forward. It is exactly the same for diving birds. They are constantly striving to go forward and down while being buoyed upwards and resisted in their progress: it must be like trying to escape a maniac in a dream.

Big birds have the strength but need perfect streamlining and small, blade-like wings to work properly when submerged: hence the penguins, which fly underwater but not through the air. To fly in both air and water, only small birds work efficiently. Even then, the size of wing, small to reduce drag and surface area so that they can be 'flapped' underwater, makes them inefficient in the air. Something like a Little Auk or a Puffin is about as good as can be devised.

Larger auks have greater problems, and the extinct Great Auk could not compromise enough to do both: it lost the power of aerial flight entirely. That it is now extinct shows just how serious the compromise became. Cormorants and Shags, on the other hand, have opted for a web-footed power source below the waves, keeping their broad wings tightly closed, and they can fly well in the open air.

The deep divers from the surface are certainly impressive. **Emperor Penguins** can reach a depth of 265 m, barely credible in view of the pressure down there and the effort needed to swim ever deeper into water that constantly tries to force the bird back up. In the North Atlantic we do not have anything quite so remarkable, but **Razorbills** are exceptionally deep divers and even the Cormorant has been recorded at depths of 37 m. Down there it can work along the seabed, poking about in weed and rocks, so it is not restricted to fish that it can see from the surface before it begins the pursuit. It can dive down and explore and come across hitherto invisible flatfish, which it then chases along the sea floor. This removes the need to look for prey from the air, as a Gannet has to, and allows Cormorants and Shags to live in the same area as Gannets and auks without competing for the same food all the time.

It is the need to survive and work well as much under the water as above it that shapes these birds. They are streamlined, almost spindle-shaped in the case of the auks; either short-winged or hugely web-footed (cormorants and shags, like gannets and pelicans, have all the toes joined by one vast web); and they are thickly feathered, helping both the streamlining and the necessary insulation against the cold, deep under Arctic waters. They have also developed a variety of bill shapes to help them catch their particular prey, on top of which superficial patterns and colours develop, sometimes only seasonally, to help them display and court a mate.

Auks are small or medium-sized, gregarious birds, the epitome of noisy, dynamic seabirds at coastal-cliff colonies. They spend much of their lives at sea, coming to land only in the breeding season, although some species remain inshore for longer and the Black Guillemot is essentially a coastal dweller all its life.

These are typical divers-from-the-surface, incapable of much more than a brief waddle on land, limited in flight to straight and low (if quite fast) progress, and unable to hover or plunge-dive from the air. They often swim with heads down, peering under for fish, and then dive and pursue their prey in open water, not going down to flush out fish from the bottom like the cormorants.

The **Guillemot** is a large, upright auk on land, long and tapered with a stout, snaky, upright or withdrawn neck at sea. It has a pointed bill but a short, square tail. Guillemots ride the waves in lines and chevrons, but often, in late summer and autumn, in twos (the male and his chick). Juveniles usually follow their parent closely, looking more hunched (and shorter-billed) while the adult looks sleek. Juveniles can easily be distinguished at first by their size and frequent calls, but it is the food-begging behaviour that later becomes the best clue as the chick becomes as large as the parent.

Guillemots feed underwater, 'flying' with their wings in pursuit of fish. In winter they spend most of their time swimming, but in the breeding season they must frequently fly between the nesting cliff and feeding grounds, while movements in the autumn often bring them within range of seawatchers on headlands. The flight is low and looks fast, although something of a struggle. The long, heavy, cylindrical body is suspended below stiffly outstretched wings, which have a rapid, almost whirring, action (although not so fast as in the smaller species). Gliding is infrequent and always brief and, although Guillemots often roll over to one side, there is none of the shearing and careening of shearwaters, nor the erratic manoeuvrability of storm-petrels. They lack the ability to do much more than fly straight, or in long arcs if they want to change direction, and the overriding impression is of constant buzzing flight without a break. Settling on the sea is simple: they just drop in with a splash.

At the breeding cliffs Guillemots form large rafts on the water just offshore, frequently mixed with other auks but in most places forming the majority. They sometimes skitter across the water, or fly a little way before diving in again in a slanting plunge that goes straight on underwater, leaving a trail of foam and bubbles. Sometimes Guillemots splash and bathe in large groups with evident pleasure and excitement.

They fly from the cliff with a steep drop, often with their legs stretched out behind and to one side, levelling out until they build up sufficient speed to begin normal flight low over the sea. On their return to land they speed in towards the rocks and sweep upwards. Gravity reduces their momentum, and they judge the arrival at their own particular ledge to perfection, just reaching it and 'stepping ashore' before they come to a stop.

The **Razorbill** is similar to a Guillemot in general form but more heavily built, with a bull neck and thick bill. Whereas a Guillemot is pointed at the front, square at the back, a Razorbill is blunt at the front and pointed at the back, its longer tail often held clear of the water as it swims. Both species swim with head and neck withdrawn, but this is, if anything, more characteristic of the flatter-headed Razorbill, which looks particularly dumpy and squat on the water. A Guillemot has a snaky look, its forehead angled above the bill, the nape round and smooth; the Razorbill has less forehead, the bill and crown more aligned, and the nape is more angular.

In flight it is very like a Guillemot, its wings narrow, the leading edge held quite straight and the trailing edge kinked behind the wrist and curving forwards at the tip. If anything it has longer wings, their tips more pointed, giving a more confident flight. Like the Guillemot, the Razorbill simply flies into the water when it wants to stop, maybe bouncing once or twice before settling down.

A Razorbill leaves its nesting ledge (or cavity) with a delicate, graceful flight, swooping down on tapered wings, both head and tail depressed and wingbeats reduced to shallow flickers, changing to a powerful flight with long, deep, slow wingbeats (the wings rising high over the back) before the usual fast whirr. It is a performance quite unlike the normal low flight over the sea and one that indicates more power and ability than one might associate with an auk.

Even more than the Guillemot, the Razorbill is likely to be seen inshore in winter, and even in sheltered, almost enclosed waters, such as the head of a Scottish sea loch, during the summer.

The auks show a variation on the theme of stout bodies, short wings and black and white plumages. Here we see a winter Razorbill (top) and Black Guillemot on the water, while the Guillemot and Puffin (bottom) show the typical auk action underwater, swimming with legs tucked in and wings providing all the power.

Guillemots are abundant in some British and Irish waters, whereas Razorbills are widespread but markedly less numerous. To see a **Brünnich's Guillemot**, you have to travel north: it is exceedingly rare in Britain (and most records refer to birds found dead). It is essentially a fractionally larger version of the Guillemot, with a thicker bill, but the bill is thick-based and pointed, unlike the square-tipped, blade-like beak of a Razorbill. In autumn and winter the smaller-billed young birds, especially Razorbills, cause potential confusion.

Puffins are gorgeous birds. Anyone could spend an hour watching Puffins; some of us could spend weeks looking at them and enjoying their engaging antics and remarkable facial decorations.

On land they are more sprightly than other auks, their legs short but their bodies raised more clear of the ground so that they can walk and waddle about quite well. They look less upright than a Guillemot, rounder-bodied and with bigger, rounder heads and short, squared tails. Typically they bob their heads, twist and turn and have an almost owl-like inquisitiveness on the nesting slopes and cliffs.

At sea, they look remarkably small, too fragile, surely, to survive the rigours of winter on an open ocean. Yet that, of course, is what

Razorbills and Guillemots are close relatives and each dives from the surface of the sea to fish. Their different bill shapes, however, indicate slightly different fishing techniques and food preferences.

they do. In winter they are often farther out to sea than the other auks, true oceanic wanderers. Their horizon extends no farther than the crest of the next advancing wave.

They are round, buoyant birds on the surface, bobbing about like corks and frequently diving under to fish. The catch is usually brought to the surface and rows of little gleaming fish are carried back, held crosswise in the remarkable bill, to feed the chick.

In flight they are fast, rolling from side to side, but direct, with no fancy aerial tricks or manoeuvrings. Their wings are small: narrow and short, with blunt tips. To keep them aloft they have to beat rapidly, with much more of a whirr than a visible action, and no glides to speak of.

At a breeding site, they are likely to whirl around in great waves, using a slower wingbeat and downward glide when launching themselves from the land, but individually they can perform well in the updraughts at the edge of a cliff. A Puffin hanging motionless in the air, wings angled back, legs drooped, perched on the wind roaring through a cleft in the clifftop, like a ball on top of a fountain, is a remarkable and beautiful sight.

Even smaller and rounder than a Puffin, the **Little Auk** is abundant in the north but a rarer bird in UK waters. Usually it is seen in northerly gales on North Sea coasts, being rare in the south and west of the country, but now and then some are driven inland by storms. Once on land, or on a reservoir, a Little Auk's chances of survival seem slight. Sometimes they hang around for a few days, but usually they disappear, probably picked off by gulls or crows, or simply washed up exhausted to die ashore.

Flying Little Auks are really tiny, almost wader-like, when first seen in small groups dashing over breaking waves offshore in a November gale. They fly fast and direct, with some sideways rolling, on narrow, whirring wings which appear to have a faster flicker or vibrant effect superimposed on a basically slower action. When seen alongside flying wildfowl such as scoters or other seabirds, they are quickly overtaken and the flight speed is somewhat illusory.

As they have very short bodies and almost no neck, the wing length in proportion is not so short as might be imagined. The body shape is a hunched oval, blunt at both ends and beetle-like; the narrow wings are less flickery than, say, a Dunlin's or a Knot's, giving a more solid appearance.

When a Little Auk settles on the sea it may come in low, rearing up almost vertically, head forward, legs pushed out and wings flickering to brake its speed, before it finally flops down, rather like a Snipe settling in a marsh. On the water an exhausted Little Auk is hunched and dejected, its neck withdrawn and bill so small as to be insignificant, creating a frog-like facial appearance. A fit bird, however,

lifts its head up, showing more neck, and looks lively and active. It often swims with its wingtips drooped and tail cocked, the undertail-coverts raised and spread like a tiny feather fan, and dives under frequently, with a flourish of the wings and a tiny splash of spray.

Most auks are cliff-nesters or, in the case of the Puffin, burrow-nesters on slopes above cliffs. The **Black Guillemot** likes tumbled boulders and broken debris at the foot of cliffs or around the edges of small, rocky islands, often in sheltered waters such as the indented western coast of Scotland. It is widespread in the northern and western isles, much scarcer south to North Wales and a rare non-breeding visitor around the coasts of England.

Black Guillemots are neat, shapely auks. The head and bill are quite large, the bill tapered to a point beyond a long, pointed face, and the neck rather long and thick when held erect, with a conical taper. The body is oval, on very short legs. On land a Black Guillemot settles back on to its tarsi, body upright and penguin-like, or sits forward on its belly; either way, it is more or less immobile.

On the water it is also a rounded, buoyant bird, its tail held low along the water, its head raised. It looks smaller, shorter and rounder than a Guillemot, and much calmer and more subdued than a Puffin. In fact, Black Guillemots always look comfortable and relaxed, happy in the moderate swell of a sheltered loch or northern harbour.

Two Puffins are joined by winter-plumaged Little Auks on the sea, while a Black Guillemot passes by in the background. Little Auks have strangely frog-like expressions, with broad heads and short, conical beaks.

Black Guillemots are barrel-shaped, their oval bodies heavy and short, with very little tail but a more protruding head and neck. The wings are less narrow and straight-edged than those of a Puffin or Razorbill, with a more curved, oval or leaf-like shape with a pointed tip. There is little to say about its flight, except that it is typical of an auk: low, direct, heavy and whirring, an impression exaggerated by the broad white patches both above and below on each wing. Rarely is a Black Guillemot seen more than a metre or two above the waves.

Fishing Black Guillemots swim slowly about, seemingly taken wherever the waves happen to go, often dipping the head and bill into the water. When they spot a suitable fish, they dive smoothly under, returning, if successful, with the silvery prey held lengthwise in the bill. A bout of diving is often followed by a preen and a typical auk action, rising up in the water with a few quick beats of the wings.

Terns and gannets are largely white. This helps other birds of their own species to see those which have found food, so they can home in on a good fishing spot. It also means that, against the sky, they are less obvious because they are pale (although this is a rather dubious notion in some circumstances, as even a white bird can look practically black against a bright sky). The idea is that the fish they are preying on do not see them coming, so they do not take avoiding action until it is too late.

Cormorants and Shags are nearly black. They do not dive from the air, but from the surface as they swim. They do not feed so much on fish that swim freely about in the upper layers of water, either, but tend instead to go deeper and search for fish on or near the seabed. Their black plumage may then even be an aid to fishing rather than a hindrance: the fish do see them coming, attempt to take avoiding action and thereby give themselves away. The Cormorant flushes out fish which, had they simply remained still, might have got away with it.

The **Cormorant** is a large bird, the size of a big goose and bigger than any auk. It looks heavy and somewhat cumbersome. Its feet are big, broadly webbed across all four toes, and its bill is very strong and wickedly hooked, to hang on to a slippery, hard-scaled fish. Unlike some of the fish-eating ducks, however, it has not developed serrated, or 'toothed', edges to the mandibles. The head is large, but the depth of the bill is such that the forehead is flat; the neck is sinuous but thick. Cormorants have large, broad tails and long, angled, blunt-tipped wings.

Like herons, they are thought of as poor fliers, but they put paid to such simple theories when they are around their breeding sites, especially if they nest in trees. They can soar and glide and dive down in a twisting descent, then sweep around in broad curves to reach the nest with a final upward flourish. At inland colonies (which are few) and roost sites, groups of birds often arrive at a considerable height and indulge in these wild acrobatic descents, putting on a spectacular show.

Cormorants fly low over short distances but tend to be higher when properly on the move. They can be seen flying by at 10 or 20 m or more above the sea, frequently in lines or 'V's, with a strong, purposeful action with deep, goose-like (but quicker) wingbeats. The head sticks out, neck slightly kinked but not fully withdrawn, and the tail is equally long and full, unlike that of a diver or goose (which might be mistaken for a Cormorant). A Cormorant settling on the water comes down in a long glide, finally rearing up, feet forward, wings beating in a rowing motion, until it drops into the waves.

Cormorants swim low, sometimes all awash except for the head, although usually a good length of back and tail is seen close to. The head is tilted upwards. They frequently look into the water with the head submerged to above the eyes, then roll forward in a deliberate, slow dive. They can occasionally take a header with a clear leap from the water, although this is rarely so pronounced as with Shags.

In summer, Black Guillemots are smoky-black with broad, pure white wing patches. Their red legs may catch the eye and, if they call, they show bright red inside the mouth, too, clearly a useful aid to communication.

Shags are smaller than Cormorants, but this may not always be obvious. Even side by side on the water the difference may be hard to judge (although equally it can be easy to see); it tends to be more obvious in the air. The head and bill of a Shag are smaller, more slender, and the thin bill and rounder head give a steeper forehead shape, often a good clue to young birds, whose plumage is less distinct than that of the adults. The neck is slightly more snaky; the tail a little narrower; the whole appearance more sleek.

In flight Shags tend to move much lower, skimming the waves, and with a quicker, snappier wing action. They glide less than Cormorants and move about in ones and twos or larger, irregular packs. Both species have an obvious tendency, even if there are only two of them together, to fly close to one another rather than even a few metres apart.

On the water, Shags also tilt their heads upwards so that the bill points well above horizontal. They look shorter, rounder birds than Cormorants but are equally variable in the amount of body showing above water. Typically Shags dive more quickly than Cormorants, often with a short forward leap that takes them fully clear of the

The Cormorant (front and top left) is a large, blackish seabird, equally at home on fresh water and in sluggish estuaries. The Shag (top right and centre) is a smaller, greener relative, with a round head and slimmer bill on a more sinuous neck.

water in a short arc. Where food and Shags are abundant, large flocks gather to feed in superb rafts that often dive in unison.

Cormorants are birds of estuaries, harbours and inshore waters; Shags are found more often on rocky coasts and beneath seabird cliffs, revelling in rough water close to dangerous shores. There is a great deal of overlap, however, and habitat is of little use in identification.

In the USA, **Double-crested Cormorants** are common coastal birds. They look intermediate between Cormorant and Shag, with a less stout bill than a Cormorant but a bigger one, and a thicker neck, than a Shag. In flight, the neck is held a little more kinked than is usual with Cormorant or Shag, but this is a subtle (and variable) feature at best. They are extremely rare visitors to Europe, only recently satisfactorily identified, but unlikely to be more than irregular vagrants.

Skuas

The magic of skuas is not just all about themselves; it is also about the effect they have on other seabirds. The drama and excitement of a chase when Great Skuas take on a Gannet, or Pomarine Skuas home in on a Kittiwake, adds greatly to the unique appeal of the skuas themselves. They are very charismatic birds, but their behaviour and the panic they spread among other seabirds takes them on to a different level. Their presence generates an energy and excitement that is almost tangible.

Skuas are kleptoparasites: they steal food from other birds, but, they also catch their own food a great deal outside the breeding season, and they are not simply thieving vagabonds.

The most familiar and in many ways archetypal skua is the **Arctic Skua**. It is not a common bird in terms of sheer numbers, although it is widely seen in spring, late summer and autumn on many British and Irish coasts. It breeds only in the far north and north-west of Britain but disperses around many coasts afterwards, before heading far south for the winter.

The Arctic Skua has one of nature's most perfect bird shapes. The body is sleek and tapered, muscular about the chest but long towards the tail, beautifully streamlined. The wings are quite long and slender, angled at the wrist and then tapered to a sharp point: falcon-like is not quite a true description, as there is more flexibility, more curve, more elasticity about a skua than a falcon. The tail of the adult has a central spike, a neat point that is sometimes long and eyecatching, but less so than the longest points of adult Long-tailed Skuas. The whole demeanour of the Arctic Skua is of a bird that means business and ought not to be taken lightly.

On its breeding grounds it is aggressive to human intruders, diving from front and back, sometimes striking with its feet, breast or wings. In display it rises to a great height, on slowly beating wings, then dives down in terrific stoops, wings ripping loudly through the air, before shooting upwards again in an impressive performance. This roller-coaster display is breathtaking, accompanied by the wildest and most evocative of calls. These are wonderful birds, always uplifting to see.

Out at sea, Arctic Skuas have an easy, relaxed action, less laboured than that of the rather larger Pomarine Skua. There are short periods of regular, rhythmic wingbeats, between glides with sudden changes of tack and sideways rolls into wave troughs. If the wind is strong, the glides become more dominant, with climbs and descents on one wingtip like a shearwater.

If a gull or tern is spotted with fish, the skua (or a pair) will accelerate from a long distance, using deep, fast wingbeats that instantly convey its intent. It closes almost to touch its intended victim, instantly entering a series of twists and swoops in its attempt to force the bird to drop or disgorge its food. It is very acrobatic, remarkably fast and the most persistent of the skuas, sometimes chasing for half a minute or more before taking the dropped fish or gliding away to settle on the sea to rest. Arctic Skuas prefer to chase

Arctic Skuas may be seen in pale forms, like this, or as much darker, brown birds that look almost black when far out over a gleaming sea. The tail spike may be reminiscent of a Long-tailed Skua, but the partial breastband and large white wing flashes rule out that species.

terns and Kittiwakes, using their ability to outfly them, but they do not attack Gannets and they have very little success against auks. In general, the higher the victim above the sea when the skua attacks, the more skuas are involved in the chase and the poorer the visibility, the greater the chance of success for the skuas.

Pomarine Skuas are generally scarcer than Arctics, although on UK coasts there are times (especially late in the autumn and in spring) when Pomarines are briefly the commonest skuas. These are bigger birds, heavily built, and frequently described as deep-chested, although sometimes the whole body looks large and barrel-shaped and the curve of the chest is not so pronounced. The weight of the body looks concentrated in the middle (whereas on an Arctic it looks farther forward, around the chest). The Pomarine has a Herring

Gull-like appearance to the head and bill and broader-based wings than an Arctic. The inner wing is consistently broad and parallel-sided, whereas the tip becomes narrow and strongly tapered (Arctics' wings are more evenly slender throughout).

In steady flight Pomarines are quite slow, but powerful, more gull-like than Arctics, with fewer and shorter glides. Even in a good breeze there are rhythmic, mechanical wingbeats, and only in a strong gale does a Pomarine regularly shear.

Pomarines more frequently attack other seabirds in order to strike and kill them; piratical chases are shorter, more laboured and less successful than those of Arctic Skuas.

Long-tailed Skuas are rare birds, smaller than either of these two, but quite chesty in build, with relatively short, stout bills. The full breast quickly tapers to a slim, flat belly and long, narrow rear body, giving the most elegant, slimline look of all the smaller skuas. It is particularly long-winged, the outer part especially being long-drawn and tapered, even more so than on a slender Arctic Skua. Adults have unique, very long, flexible central tail feathers.

In normal flight, Long-tailed Skuas are tern-like and buoyant with a noticeable movement of the body, which rises and falls with each wingbeat. In a gale, the Long-tailed employs longer spells of shearwater-like glides than any of the others.

The Long-tailed Skua is also more varied in its ability to chase other birds (its pursuit of terns being much as in Arctic), to dive from the surface, and even to soar about catching flying ants. Some come to land and feed in fields, more like a Common Gull than an Arctic Skua. On the breeding grounds Long-tailed Skuas rely more on catching lemmings and other prey and less on piracy than Arctics, which take 90 per cent of their food by piracy when breeding.

Frequent in the northern isles and around some coasts and headlands in autumn, **Great Skuas** are massive birds, bigger than even the biggest Pomarines. They have particularly wide bodies and broad wings; the outer wing is fairly short in proportion, triangular and pointed. The tail is short and wedge-shaped or square, with no real projection.

Great Skuas ashore are more mobile than the short-legged, long-bodied Arctics. They attack with vigour and skill, powering in to an intruder at chest height and then swooping up at the last minute in a truly hair-raising performance that may end in a strike to the head. They also soar above their territories like big birds of prey.

At sea they fly low, frequently sidling down into deep troughs and rolling up over the next crest. They have regular, quite quick wing-beats, although, because of their size, they tend to look relatively slow and heavy. The wingbeats have a noticeable upstroke with bowed wings, tips angled down. In a chase they are clearly less agile

Great Skuas are big, almost cumbersome, but extremely powerful birds (top right), while Long-tailed Skuas (centre) are elegant, almost tern-like. The Pomarine Skua (bottom) looks more like a heavy Arctic Skua with a blunt, spoon-like tail spike.

and lightweight than an Arctic, or even a Pomarine, but they make up for this with superior power and a willingness to take on bigger opposition, even tweaking the wings of Gannets or hitting them with their feet and forcing them into the sea. They prefer to take their victims by surprise, rather than using their speed to catch up with them, frequently soaring at a height while looking for potential prey and then diving at speed.

Great Skuas, or Bonxies as they are often called, are more predatory than the smaller species, hitting gulls and other birds chest-first, with their bills or using their feet, and knocking them to the water. The hapless victim is then hammered by the skua's stout, hooked bill and disembowelled as it floats upside-down.

Great Skuas are frequently in attendance at trawlers and in the wake of large ships, where they often dominate the gulls and Fulmars that are also trying to take the richest pickings (and which, in turn, may dominate the smaller skuas). Great Skuas have a particularly mean, assertive character, instantly separating them from the superficially similar brown juvenile gulls, and form an impressive minority in the mêlée.

Rarely reported (and usually not 100 per cent identifiable) in the UK, but a little more frequent in the western Atlantic, the **South Polar Skua** is a bird that dedicated seawatchers hope to encounter in autumn gales. These huge seabirds look like Great Skuas, sometimes more like large Pomarines, and dark-phase individuals are extremely difficult to separate from occasional all-dark juvenile Greats. Compared with some of the mammoth southern Great Skuas they are fractionally smaller-billed and more slender-winged, but compared with many Great Skuas from the North Atlantic there is very little to choose between them in size. Bill size is very similar (certainly too close to rely on in the field), and even the breadth of the wing is hardly much different, although smaller South Polars may look noticeably slim. In general, they are remarkably like Great Skuas, but pale-phase adults can safely be separated on plumage features given a close view.

Piracy is used by many other seabirds from time to time, but not so importantly as by skuas. Many gulls chase other birds, often other gulls, to try to steal food, and even Common Terns will do so, but the biggest pirates of the lot are the frigatebirds. Frigatebirds are practically unknown in the eastern Atlantic, so rare as to be ignored, but the **Magnificent Frigatebird** breeds off southern Florida and occasionally wanders farther up the eastern coast of the USA.

Frigatebirds are among the most striking, rakish birds of all in the air. They have very long, narrow tails which may be held in a single, thin spike but open up to show a deep fork when spread. Their wings are remarkable: long, narrow, pointed and angular, held in a flattened 'M'-shape. A frigatebird cruising and intent on a meal flies quite high over the sea, usually gliding or maybe giving just the occasional slow, deep flap in slow-cruise mode. It looks a little like a giant Red Kite as it soars in flat circles and dives down to a likely meal with a sudden flourish, but its wings and tail are all tapered to much longer, more extravagant points. When it needs to step on it, to chase a tern or a tropicbird, it can do so with impressive ease, becoming exceptionally fast and agile, its wings and tail now allowing a wonderful degree of manoeuvrability.

MIGRATION

One of the many things that make seabirds so magical is their ability to migrate vast distances over open oceans. Migration is far from unique to seabirds: indeed, there are very many species, and thousands of millions of individual birds annually, that perform long-distance migrations, every one a minor miracle. Yet seabirds seem extra special, simply because they can do it over totally trackless seas, literally without a landmark for many thousands of kilometres. The only thing going for them is that, if they get tired or hungry, they can settle on the sea, or snatch a fish or squid as they go, an option not open to a songbird crossing an ocean or a desert.

Quite how a tiny storm-petrel or shearwater, or even an albatross (which, in the context of the sea, is still minute), can spend years out of sight of land but still return to its place of birth is beyond all of us (and even finding its own burrow, in the dark, in a colony of maybe 100,000, is a staggering achievement). It is not too difficult to understand how a bird can decide which way is north and judge where to head if it must fly south, or west: it uses the position of the sun, stars and moon (probably reacting to polarized light so that it can do so even in overcast conditions), the earth's magnetic field and other clues. The miracle is that it knows in which direction to travel: turning due east (or any other bearing) is no problem, but knowing that it is due west of where it intends to be a few days hence is really clever.

Most humans have some vague sense of direction and, given a clear night, or a view of the sun and a clock, we can gauge roughly where north is and work the rest out from there. We can know that we must travel on a particular bearing to reach a certain point, however, only if we know where we are, and have a map that shows the relative position of the destination. Without landmarks or signposts to help, we are lost.

Put anyone in a boat somewhere in the middle of the North Sea, out of sight of land, and say 'Head for Aberdeen', and it could not be done. Is Aberdeen north-west, west, or south-west? Or, worse still, is a particular oil rig, for example, also out of sight of land, to the east, west, north or south of us?

This is the wonder of bird migration: not so much the navigation, but knowing which way is home and being able to fly with such precision that, from a viewpoint just inches above breaking waves or even deep down inside wave troughs, the bird can reach its

Storm-petrels are mysterious and magical birds; here a Leach's Storm-petrel feeds at sea in the light of a full moon. Dark night in summer see the petrels return to their burrows on remote, rocky islands where they would be in danger from gulls by day.

own special island. Yet **Manx Shearwaters** have been carried away in boxes from South Wales, to the coast of Brazil and to the Mediterranean, and, without any way known to us of knowing where they are, released to fly home. And fly home they do, within a matter of days.

Actually, thinking of seabird flight over the sea from a human point of view, even the everyday fishing trip takes on a new significance. A **Puffin**, say, far from land, catches a few fish and decides to fly back to base. It skims along above the sea, probably in wind and rain or sea mist, in and out of wave troughs and crests that reduce visibility to a few metres at most, for most of the journey. Yet, with no particular difficulty, it makes a beeline for its burrow on a low-lying island or piece of coastal cliff.

The point of migration is that it takes birds from a place that will shortly be unable to support them to an area where they may survive adequately for the next season or so. The reason behind the move may be climatic: northern species may have to go because the sea is about to freeze. It may be to do with food: movements of fish may require a readjustment as times of plenty threaten to become times of famine until next spring.

Manx Shearwaters breed in the North Atlantic while Sooty Shearwaters (right) breed far to the south and come north in their non-breeding season migrations and are present in European waters during our summer.

Some auks move far from land because they are moulting, and lose the power of flight for a while. They literally swim the bulk of their migration or dispersal routes, perhaps followed by chicks for several hundreds of kilometres. Moving away from land, to give themselves room, so that they can ride the winds and waves, allows them a degree of freedom that too close a proximity to the coast would deny. Moulting birds caught in onshore winds for long spells too often end up as wrecks – whole flocks of auks driven ashore, exhausted and starving. It is far better if they are able to live in the open sea without the constraints of a nearby coast.

Many terns and skuas from the coasts of Britain and Ireland migrate quickly south, through Biscay to northern Iberia, along the west coast of Portugal and on to North Africa. Then they reach the bulge of West Africa or continue through to the Gulf of Guinea, where their movements are related to the abundance of small fish, principally Sardines. Manx Shearwaters, on the other hand, move more westwards, heading across the Atlantic and down to waters off the coast of Brazil. Gannets are more randomly dispersed, but young ones move farther south than adults, heading for north-west African waters. There they may stay for a year or two, before going back to base: an island which they saw from the sea only during the few days it took them to escape from it by swimming away.

Some southern seabirds, such as the Great Shearwater and Sooty Shearwater, have long but regular routes that take them in huge circles or figures of eight, sweeping northwards up one side of the Atlantic, across the top, then down again on the other. Such routes are even more extensive in the wider expanses of the Pacific.

Great Shearwaters start breeding on Tristan da Cunha, in the mid Atlantic opposite the coast of northern Argentina, in November. In late April and May they move north, crossing less rich, warmer waters at first, moving slightly west of north to reach the North Atlantic off Newfoundland. As the northern summer progresses, they swing eastwards across the Atlantic, sometimes getting close to Irish coasts but often falling short before curling south towards the Azores and the Canaries. They pass close to West Africa and then, cutting just west of south, reach Brazilian seas, completing a figure of eight by returning east to their breeding islands again.

All this way they have adequate food supplies and they are also using the winds: they have good tail winds or cross winds that allow them to employ their specially dramatic, dynamic shearing flight without burning up much energy in all their prodigious travels.

The **Arctic Tern** is often acclaimed as the greatest traveller of all, with the cliché that it sees more daylight than any other bird as it spends the summer in the land of the midnight sun of the Arctic and another summer, with yet more midnight sun, in the Antarctic. It does, indeed, have a remarkable life, during which it travels vast distances from the islands of the far north to the southernmost part of the Atlantic and even around Antarctica, using the roaring westerly winds. The period spent in the far south, where the waters are rich in fish, is comparatively lengthy and leisurely, while movements south and back northwards through the tropics, which are less productive, are much faster.

Shorter distances are covered by some of the auks. British and Irish **Razorbills**, for example, move south-west in the autumn, to Biscay and northern Iberia. Young ones move on later, often entering the Mediterranean and reaching as far as northern Italy. Adults are less dispersive, tending to remain in the North Sea, the Irish

Sea, the English Channel or the Bay of Biscay and much more rarely getting to Mediterranean waters. Kittiwakes, Shags, Fulmars and the Gannets already mentioned are other examples of species in which immature birds move much farther than adults.

Simply watching gulls at a coastal or inland roost during a few months shows the differences in timing and movements of adults and young. Many winter flocks in southern Britain consist largely of adult birds; by late winter most have gone. In spring there will be a surge of more of the same species, but now largely browner, young birds that have wintered far to the south and have no need to move back to breeding colonies early in the year.

That things sometimes go wrong has been illustrated by the occasional vagrants which appear at about the right latitude for the time of the year but in quite the wrong ocean. An Aleutian Tern in north-east England and an Ancient Murrelet in the Bristol Channel are two famous examples: both should have been in the Pacific, but

Arctic Terns breed right up into the Arctic Circle but in the northern winter they are down in the southern hemisphere, flying among the icebergs of the Antarctic. They perform some of the longest migrations in the bird world.

Longish legs (for a tern), a pink flush to the white breast and a largely black bill make the spring Roseate Tern a distinctive rarity of great beauty. For reasons that are still unclear, it has undergone a disastrous decline in numbers in recent decades.

had somehow made a mistaken crossing around the north of the Americas or around Cape Horn before resuming their north–south movements as before.

It is the species that spend months far from land during the winter, species as diverse as Puffins, European Storm-petrels, Sabine's Gulls and Long-tailed Skuas, that evoke the greatest admiration. They have tackled the sea at its toughest and won. To survive at all in such conditions is just remarkable.

BREEDING

Once a seabird is mature it must come to land to breed. No bird is truly a halcyon, able to lay its eggs at sea and calm the waves until they are hatched. All need a solid base at which to court a mate, make a nest or claim a site, lay eggs, and feed chicks until they are at least old enough to swim out to sea.

The odd thing about many seabirds is the length of time needed to carry out the whole process. First of all they take several years to mature (especially the albatrosses, large shearwaters and Fulmars). They may spend years visiting likely sites before plumping for a ledge and deciding on a mate. Fulmars even sit about on ledges for several summers before getting down to laying an egg.

Then the amount of time required simply to get through the breeding process is very long, even for the 'tinies' like the storm-petrels. For the big albatrosses it is so long that one breeding season begins to impinge on the next year, and the birds have to take a year off before the next round begins: they cannot squeeze in an annual breeding attempt.

One reason suggested for the long period of immaturity of seabirds is that the colony, necessarily concentrated on a small area of suitable land (a cliff-bound islet for instance), is vulnerable, to disease, storms or other such disasters (nowadays pollution and overfishing can be added, but the birds may not have evolved with massive oil slicks or nylon nets in mind). To have a great number of widely dispersed immatures at sea when the breeding adults are overtaken by disaster is a good insurance policy: there will still be birds to come to breed in future years.

Another is that it takes years of experience to come to terms with the demands of life at sea. An immature bird may have its work cut out to feed itself for a long time, until it learns the ropes and becomes more proficient at finding food and catching it. When it begins to breed it may have to feed itself, feed its mate (to help build the heavy, energy-packed eggs) and then also feed its voracious chicks, and to do so it has to be good at the job. A good indication that this is correct may be the difference in distribution of adult and immature birds of some species: young Gannets, perhaps, find life easier at sea in the warm waters off Africa, while the adults do not need to go for such easy pickings in later life. Moreover, older, experienced breeding pairs of some species, such as Common Terns, certainly produce, on average, more young per breeding attempt than do young pairs.

Seabirds actually occupy a number of different habitat types when breeding. Not all, by any means, live down a hole or nest on a ledge above a dizzy drop into the sea. The dramatic ones, though, breed as close to the edge of the sea as they can get, and a great seabird city is one of the joys of nature for any wildlife-watcher. A vast penguin rookery in the Antarctic, a great gannetry in the North Atlantic, a mass of Kittiwakes and Guillemots on a North Sea crag, all are, in their own ways, sensational. It is a combination of sound, smell, movement and vitality, plus the strange attraction of great numbers of identical elements in a scene (like ants in a nest or bees in a hive, or a field of dancing buttercups), that gives it a special appeal.

Other species, however, choose to live differently. Many terns nest on flat beaches or lumpy saltmarshes. Gulls sometimes take themselves inland to a heathery moor, or prefer rocky outcrops at the edge of a beach. Shearwaters like burrows on top of smooth-profiled islands, as well as holes in clifftops, and even cavities in the tumbled scree of a mountainside. Storm-petrels find their way home in the dark to the dry-stone walls of ancient brochs and hermits' cells. Kittiwakes have even decided to make their home on the windowledges of a flour mill in northern England.

BREEDING BEHAVIOUR

The great majority of seabird species are gregarious when breeding, nesting in colonies that are often large and usually dense. Although only about 15 per cent of the world's birds nest in colonies, 95 per cent of the seabirds do so. They create, incidentally, some of the finest spectacles in all birdlife, matched only, perhaps, by the flamingos of the Great Rift valley and some of the awesome winter wildfowl flocks of Europe and North America.

In British and Irish waters there are large colonies of several species. On Skomer Island, off the coast of Dyfed, over 100,000 pairs of Manx Shearwaters occupy burrows across the island top. On the islands of St Kilda, west of the Outer Hebrides, some 50,000 pairs of Gannets breed, as well as perhaps 250,000 Puffins. Grassholm, another Dyfed island (not much more than an islet), has around 28,000–30,000 Gannet pairs, a population apparently stable by the beginning of the 1990s after a period of rapid growth. Bempton Cliffs in Humberside, just north of Flamborough Head, have scores of thousands of Kittiwakes and Guillemots. Orkney is full of exciting seabird cliffs with magnificent colonies.

At each colony, the seabirds of any one species behave in some way that creates a degree of social interaction and stimulus. This ties them together into what may properly be called a colony rather than a random collection of birds that happen to be in close proximity. Bryan Nelson, trying to define a colony, considered that it is 'a group of potentially interbreeding individuals, each of which is influenced by being part of the group.' It is the fact of being in a group and therefore behaving, or succeeding, in a way that would not otherwise have happened that is important.

The social and individual benefits that accrue from being part of a colony are several and varied. It is likely, but hard to prove, that birds learn the whereabouts of good feeding grounds by following others in the colony on fishing trips (rather like honey-bees leading others to flowery meadows, and Starlings following well-fed fellows from their winter roosts). This could, however, be the case for some species and not others, and even, conceivably, for some colonies at certain times, but not at others.

Colonies certainly allow birds to react more quickly and more effectively to predators. One tern can do little to deter a marauding fox (or even a hedgehog, come to that), but a flock of a hundred or two raining blows on to a detested intruder has a greater deterrent effect. An individual bird is also likely to be caught and killed by a predator if it is discovered alone, because there are no options. Simple rules of chance show that in a group there is a greater likelihood of survival: the predator is more likely to 'kill someone else'. It benefits the one that is killed not at all, and the population still declines by one, but, individually, the chances of not being that one are better.

Social activities, such as display, aggression, noisy calling and courtship-feeding, stimulate and excite seabirds, which helps them to come into breeding condition at more or less the same time. When a pair of **Common Terns** begins to court and mate, it may set off a whole chain reaction of other pairs doing the same thing as if unable to contain their excitement. This increased synchronization is beneficial to most species in their breeding, although some still retain long seasons within a colony as a whole, such as the **Gannets** which still have small downy chicks after the older juveniles have fledged. In some species, such as Common Tern and Gannet, sub-groups within the colony are closely synchronized, even if the colony as a whole may not be.

Bryan Nelson also argues eloquently that a colony is safer for eggs and chicks if all the birds are going about the same sort of business together than if they are working at random. **Herring Gulls** may take other Herring Gulls' eggs and chicks, but if all are sitting on eggs or busy feeding young they will have less time and opportunity to cause mischief within their own colony. In colonies that are very large, such as some Common Tern colonies, there are opportunities for those sub-groups within them to form. These are closely synchronized, even if other groups are somewhat ahead of or behind schedule. This means that at least the birds that are nesting side by side are pretty well together, even if others from the other side of the island are out of phase.

It is probable, too, that a glut of young at once is better than a constant trickle of chicks spread out over a longer period. Predators can catch, eat and feed their young on only so many young seabirds at once: better that the predation is over and done with in a short time than to allow predators a continual supply of food all summer. The latter situation would let the predators rear more of their own young, and live at a greater density, and so kill far more seabirds in total. This assumes that the general rule applies, that numbers of prey determine numbers of predators and not the other way around. It is a likely situation with seabirds but is perhaps easier to see in species such as the Barnacle Goose. The adult geese produce young in Greenland at the top of a cliff: the goslings tumble dangerously down the cliff after hatching, many to be gobbled up by Arctic Foxes; but there are so many chicks all at once and so few foxes that, eventually, a few goslings get through because the foxes cannot cope with eating any more.

There are also factors that are influenced by density and position within the colony. Among **Herring Gulls** studied in Scotland, both high and low density compared with the average density within a colony meant later nesting. The 'norm' encouraged an earlier start, and the birds at that density tended to have larger clutches of eggs and reared the most young. Birds at the edge of a colony are more likely to suffer predation.

There are all kinds of fascinating correlations of this sort. **Kittiwakes** and Herring Gulls achieve better breeding success early in the season than later. Other species survive better in the early months of their lives if they have been reared early in the season than if they have been reared late. Late-fledging **Gannets** face a difficult time going to sea in the autumn storms. Perhaps other species, hatched late, survive less well because they are fed inferior quality or quantities of food? There may well be many more intriguing things to discover given sufficient research, as has been the case with landbirds as varied as the intensively studied Great Tit, bee-eaters and Dunnock.

Colonies may allow birds to find a mate more easily, but in the case of some species, such as the **Common Tern**, pairs are often already formed before their arrival at the colony in spring. Gulls and terns may often be seen paired and performing courtship behaviour on migration in spring, and the pairs may very often involve the same individuals that were paired the previous year, reinforcing a bond that lasts for a number of seasons.

A Guillemot chick is shepherded by its father for the first few weeks of its life, having been coaxed from the cliff ledge to plunge into the sea at about eighteen days old.

The age of first breeding by seabirds has already been referred to, being noticeably later, on average, than in most landbirds. A Guillemot may breed at two or three years old, but Razorbills do not do so until they are five or six and even the smaller Puffins wait until four or five years of age. Terns mostly do not breed until four years old, and Kittiwakes at four or five (sometimes not until they are eight). Gannets, despite their size, breed at the relatively young age of four or five, but tiny Leach's Storm-petrels remain at sea for just as long before returning to breed for the first time. Fulmars may not nest until they are seven, even eleven, years old, although pairs may form and sit about on ledges for a few years before laying an egg.

Incubation of the egg, or eggs, then takes a long time, too. Albatrosses may take up to 79 days to hatch their egg; Manx Shearwaters incubate for 51 days, and the European Storm-petrel incubates for 41 days though it is barely Starling-sized (a similar-sized landbird might hatch its egg well inside a fortnight). Part of the reason for this long stint in the small Storm-petrels is that they are physically at a lower temperature than most small birds, perhaps by 3 °C: it takes longer to hatch the egg simply because it is not warmed so much.

The chicks are not especially well developed on hatching and still require feeding for an awfully long time before they fledge. The delivery of food may be erratic, and chicks must go long periods with no food and often at low temperatures while their parents are away at sea. Like young swifts, some young seabirds, especially the Storm-petrels, are able to lower their temperature and use less energy while waiting long days for the next meal.

Guillemots leave the ledge where they were hatched, accompanied by a parent, after eighteen days, but they are not then able to fly and are still only a third grown. Young Puffins, on the other hand, remain in the burrow for 46–53 days and then leave without any attention from the adults. The baby Guillemot leaves the ledge when it is small, but is fed (or, because it is fed?) by its parent until it is full grown; the young Puffin is full grown and can feed itself as soon as it

For several nights when they are almost old enough to fly for the first time, Manx Shearwater chicks peer out from their burrows. Then, one night, they scramble away to a suitable mound or cliff edge and fly off, to face several years alone at sea.

leaves, because it will not be looked after by its parents (or, its parents do not need to look after it because it feeds itself).

European Storm-petrel chicks remain in their scented, dark and eerie tunnels for 56–73 days before flying out to sea one dark night, quite alone, not to return to land for several years but then probably finding the same little island again. **Manx Shearwater** chicks, deserted by their parents, peer from their burrows for a few nights until, at 62–76 days old, they scramble out to rise and flutter away into the moonless void. Fulmar chicks do not fly for over forty or fifty days, but terns such as Common and Arctic Terns fledge in a shorter time, around 23–28 days.

Once the chick has fledged it may be entirely on its own or, in some species, may be looked after and fed by its parents for a time. Young gulls and terns fly around after their parents squealing to be fed. Common Tern chicks may even migrate to Africa with their parents and remain with them through the winter. Cormorants feed their chicks for a further five or six weeks after they can fly.

Gannets do quite the opposite, stuffing their chick with food until it is so fat that it really cannot fly, but keeping the fledging period as short as possible in the run-up to the stormy days of autumn. During the last few days of its 84–97-day stay at the nest the young Gannet may refuse food or show less interest because it appears at last to be replete, or it may be fed until the moment before it finally flies. Either way, the parent Gannets attend their chick and try to feed it until it decides that it has had enough and must face the dangers of the unknown sea, and the more immediate terrors of neighbouring Gannets, on its own.

Deserting its parents, the young Gannet scrambles away and belly-flops into the sea, having to swim out from land and starve for a while until it is light enough to take off properly for its first real flight. Then, unaccompanied and untaught, it perfects its hunting technique and flies off to Africa for a year or two in the sun where the pickings are easy, before returning north.

Cliff-nesters

A substantial proportion of our seabirds breed on (or just above) cliffs, including some that can and do nest in other situations, too. **Fulmars** occupy large, often earthy and grass-bordered ledges, near the clifftop. Grassy slopes above cliffs, where landslips and minor falls

Good seabird cliffs with mixed colonies may tend to stratify, with Fulmars near the top, Kittiwakes and Guillemots lower on tiny, narrow ledges, Razorbills in deeper crevices and hollows, and Shags and Cormorants occupying the broader ledges that can cope with their large nests.

have left hollows and cracks, make ideal Fulmar sites, but they nest all over the place where they are common, as in the Shetlands, where they occupy low banks and walls, rooftops and even flat ground inland and ledges above 'cliffs' only a metre or two high at the shore.

Fulmars use buildings, such as the walls of Bamburgh Castle in Northumberland and windowledges of Dunrobin Castle in Sutherland. In a number of areas they even breed out of sight of the sea, or away from the open sea at least, such as on rocks above The Mound, Sutherland, and as many as 20 km inland in Yorkshire. The egg is laid direct on to the earth or rock, with no real attempt at anything other than perhaps a rough scrape for the nest.

The Fulmar is famous for its ability to eject a stream of foul – (and long-lasting) – stomach oil from its mouth (and perhaps also its tubular nostrils) at intruders. These may be in the form of rock-climbers ignoring appeals to avoid seabird cliffs in the nesting season, foxes, Peregrines or even White-tailed Eagles. Some of the latter, introduced on to Fair Isle, died at the hands of Fulmars, drenched in sticky oil. Even chicks squirt oil in a powerful, well-aimed jet.

At the nest the Fulmar is not very mobile, being unable to walk or properly stand on its feet, remaining flat on its belly almost all the time. Fulmars are noisy and demonstrative birds, creating a wonderful cacophony of throaty, choking, rattling calls that echo around enclosed inlets or against gigantic sea stacks (such as, respectively, Smoo Cave and the Great Stack of Handa Island, both in Sutherland). They voice their demonic laughter with their large mouths wide open, heads held up and pulled well back, or rocked from side to side as a pair indulges in energetic, ecstatic courtship posturing.

European Storm-petrels (and Manx Shearwaters) may be found in crevices in cliffs but are not cliff-nesters proper and will be considered later. **Gannets** are, however, principally cliff-ledge breeders. At Bempton Cliffs they nest on broad limestone ledges on sheer cliffs. Bass Rock has ledge-nesters on steep cliffs, as well as others spilling out on to the top of the island. Grassholm has less sheer cliffs, but the colony spreads up from the steep island sides to the rounded top, occupying almost level ground on the upper sections. Ailsa Craig is a spectacular Gannet colony, a great, domed island ringed with Gannets on the sheer cliffs of its skirts. At St Kilda, the huge stacks of Stac Lee and Stac an Armin and the cliffs of Boreray shine white from afar, smothered with breeding Gannets for more than half the year. In the western Atlantic, Bonaventure Island is one of the most atmospheric Gannet colonies of all, its brooding cliffs sometimes looming out of thick mist, dotted with Gannet pairs.

The Fulmar is generally quiet at sea and in the air but is a noisy raucous creature on the nesting ledges, where pairs often cackle together.

Gannet nests are bulky affairs, made of seaweed and, all too often, netting which litters the sea and frequently kills both old and young Gannets, which become entangled in it and starve. Several things strike the visitor to a Gannet colony. The sky will often be thick with wheeling birds, circling and hanging in the breeze, quite magnificent in their spotless white feathers and with 2-m spans. If the day is fine, the sight of the dazzling birds against a deep blue sky is sensational. On the ground, the regularity of the largest colonies is remarkable, a thicker band of non-breeders in 'clubs' around the edges framing the breeding colony of extraordinarily uniform density. And the sound is uncanny: an endless chorus of curiously mechanical, yet excited, throaty, deep, 'ur-rah ur-rah ur-rah' calls, monotonous and endless like some giant, rhythmic machine.

On broad ledges at the top of sea stacks or high on sheer mainland cliffs, **Cormorants** make big nests of sticks, heather, waste netting and seaweed. Tree-nesting inland persists in Ireland but disappeared in Scotland long ago; now a few colonies have sprung up in eastern and central England, often beside breeding herons in treetop siuations. The trees become white with droppings in a great spray below each nest, often dying after a few years from the toxic effects of the guano.

Cormorants are big, striking birds, not everyone's idea of beautiful creatures but, in fact, undeniably handsome in early spring: beautiful, true enough. They perform strange, ritualized neck-writhing displays and stand with tail raised, wings lifted up from the body and bill stretched forward, wide open, to reveal the vivid yellow gape. The Cormorants are often seen flying about the colony carrying weed and sticks in their bills to add to the nest structures. Their voices let them down: grunts and indiscreet groans from the adults, and high, almost warbling choruses from their gruesome chicks.

Shags also breed on sea cliffs, much more exclusively so, in fact, than Cormorants. Some colonies are large and widely scattered, while others are small, tight groups on short ledges on a big cliff face. Some sites are exposed, whereas others are much more sheltered and hidden away in deep caves, under overhangs or on the landward side of stacks. Boulders that fall in chaotic piles at the foot of cliffs also provide nesting places for Shags. Like Cormorants, they soon turn the cliff face white with droppings, which spray in huge splashes below nesting and roosting sites. Old, experienced nesters claim the more secluded, sheltered sites, while younger birds often have to attempt breeding on exposed places and find their nests are washed or blown away. The nests are built of thick, heavy stalks of weed and bracken, lined with finer material on top. Disturbed birds lean forward, cackle loudly and writhe their snaky necks in an impressive threat display, a cold gleam in their gloriously green eyes.

Herring Gulls are flexible in most things they do, and many nest on cliffs while others prefer buildings, flat islands or even sand-dunes. On cliffs they tend to nest near the top, on broken ground or broad, but short, ledges. They make a variable nest of wisps of grass or substantial amounts of plant material.

Great Black-backed Gulls choose the highest sites, often on top of a pinnacle or a tall sea stack, dominating the scene with their strikingly contrasted appearance and loud, gruff calls. Herring Gulls are regularly colonial, whereas Great Black-backs are more often solitary or widely dispersed. Nevertheless, some sites, from Cornwall to the Outer Hebrides and Orkney, have substantial colonies, though the largest is still of under two thousand pairs. The nest is often sheltered by a boulder or wall on which the adults frequently stand; it is made of dry vegetation gathered into a slight hollow. The big birds behave like most gulls, although they are a little less aggressive to humans, strangely enough. They patrol their territory offshore or above the cliff in leisurely, floating glides and call loudly (and with very deep voices) from vantage points on the ground.

One of the cliff-nesters *par excellence* is the **Kittiwake**. It is the gull of the sheer sea cliff, nesting on long, sinuous ledges and also the tiniest of prominences that give a toehold for its nest. Some sites are therefore beautifully regular and linear, while others have odd nests dotted about everywhere. Green algae and droppings cement the nest foundation to the cliff, on to which mud and vegetation (much from inland) is compressed. Kittiwakes add immeasurably to the drama and atmosphere of a seabird colony with their incessant calling and the sweet, fishy smell of their droppings (which a seabird enthusiast would describe with affection, while most other visitors turn up their noses).

Kittiwakes choose both exposed cliffs and more sheltered ones, but often take the ledges on the lower half of a sheer cliff, almost down to the point where they are washed by the waves. Many sit about on rock slabs and wave-cut platforms at the foot of the cliff (and, surprisingly, they frequently resort to sandy beaches nearby in the summer). Some colonies are simply vast: islands on the fringe of the Arctic have hundreds of thousands, and in Britain there are tens of thousands in many colonies from Bempton northwards, especially in Orkney. Marwick Head on the mainland of Orkney is among the biggest of the British colonies. In such concentrations the effect is awe-inspiring, but the individual Kittiwake is no less beautiful, with its sparkling eyes, yellow bill and most intensely vermilion gape.

None of the terns is a cliff-nester, although many will be seen offshore, on stacks and on lower rocks at the foot of cliffs. Skuas, likewise, are not cliff-nesters, but they are marauders over the adjacent seas and may even take young seabirds from cliff ledges.

The **Razorbill**, however, is a cliff specialist. It likes secluded cavities and clefts, or even the sheltered corners at one end of a longer ledge, as well as holes in boulder scree much lower down. Razorbills are often, therefore, seen in ones and twos, not so often in long lines squashed shoulder to shoulder. They have beautiful display flights from the cliffs out over the sea, and frequently join Guillemots and Puffins in large concentrations swimming below the nesting cliffs.

Guillemots would rather be in dense concentrations, often in long lines on extensive horizontal or sloping ledges across the cliff face. Sandstone with deeply cut indentations makes an ideal Guillemot cliff, but others are on chalk, limestone and granite. Other colonies are in different situations, but are much less frequent: in some places Guillemots occupy the flat tops of sea stacks, jostling together like penguins at a rookery.

In spring the activity is at its height, with the birds constantly bowing and weaving on the ledges and flying in and out, hundreds more in the air and thousands on the water below, mixed with other auks. There is also a constant rising and falling chorus of whirring, trumpeting calls, vying with the Kittiwakes' noisy squealing. Birds on eggs are hunched down, with one wing slightly lowered, all in a row, faces to the cliff, shoulder to shoulder.

Puffins are often dotted about seabird cliffs in twos, threes and fours. Usually they are close to a deep crack or defile, or on the edge of a broad fan of loose or grass-grown debris in an indentation of the cliff. They also nest on long, slippery grass slopes above cliffs, or among tussocks of thrift and boulders at the cliff edge.

The typical image of a Puffin puts it on Old Red Sandstone, flanked by the tufts of pink blossom and curled leaves of thrift and backed by a blue sea with white horses out to the far horizon. Puffins, and such blissful coastal scenes, are irresistible. Who would not want to be out on a sunny, windswept isle in the Puffin season?

In reality, of course, while there are many such days of sun and clear air, there are also many of hanging, cold sea mist or driving rain and crashing salt spray. Puffins nest in burrows, often taken over from rabbits but also frequently excavated or extended by the Puffins themselves. Compared with the other auks, the Puffins are more mobile, less restricted to a narrow ledge and able to walk about on their strong, short legs, bodies held up clear of the ground.

Moorland-nesters

A number of seabirds need open space in which to nest, free of the confined ledges of sheer cliffs above the waves. They may have substantial territories, or at least wander over sizeable areas on the ground, or have tiny territories within large tight-packed colonies.

Black-headed Gulls breed on moors far inland, such as high, peaty plateaux in Mid Wales, where they occupy the rushy margins of cold, sterile tarns and sometimes become marooned on the dried-out, dusty, black peat beside tiny moorland pools which will dry up in a hot summer. The larger colonies (of several hundred pairs) are always beside a bigger lake, but there are many tiny clusters of two, three or half a dozen pairs scattered over upland regions. The birds arrive in March, instantly beginning their loud, squalling dawn choruses daily.

Common Gulls, medium-sized, gentle-faced gulls with lovely, high-pitched squealing calls, spend relatively little time at sea. In winter they often feed on pastures and playing fields, where they can stride freely over short grass, looking for worms. They roost on beaches and inshore waters, however, and are also quite at home following ships far offshore (although it is the Lesser Black-backed Gull that follows ships farthest from its home shores).

In summer they move to small colonies, often with under ten or a dozen pairs, rarely a hundred or more, anywhere from sea level to an altitude of 1000 m in Scotland and higher in Scandinavia. Small islands, sand-dunes, rocky outcrops above small beaches, shingle, saltmarsh, moors and even cereal fields are used for nesting. It is this gull that most readily nests in trees, usually dead ones with a few, large, branches, when the nest may be a more substantial structure than it often is on the ground. A typical colony, however, may be on rather damp ground with heather and cotton-grass over a soggy layer of sphagnum moss, where the pure white heads of incubating gulls make a strong contrast with the dark-toned moor.

Lesser Black-backed Gulls similarly use a variety of sites, from coastal cliffs to rooftops in cities. The colony is generally larger than a Common Gull colony, and often mixed with Herring Gulls. Compared with the Herring Gull, however, this species shows a greater tendency to nest inland and prefers, on the whole, a more level tract of ground. This tendency is clear even where the differences between level and sloping are subtle, as on Walney Island where the Lesser Black-backs nest on the flat centre while Herring Gulls occupy rough ground around the edge. There are some very large colonies, of thousands of pairs, on northern moors inland, on heathery ground; the coastal island sites tend to be in marram, bluebells and mixed maritime vegetation such as thrift and sea campion.

Lesser Black-backs have much the same courtship and territorial behaviour as other large gull species, with a great emphasis placed on loud calling in special forward-leaning postures on the ground. Towards human intruders they are bold and aggressive, calling with sharp, staccato notes and diving repeatedly at people's heads, although with less venom, or at least less style, than the skuas.

Elegant, dark-eyed Common Gulls nest in some remote places on damp, heathery moors and mires, as well as on rocks beside the coast.

Arctic Terns are regularly moorland-nesters in the northern isles of Britain. They nest on many rocky-shoreline sites and storm beaches with large stones piled up at the head of a windswept bay, but there are large (though fluctuating) colonies in Orkney based on bleak maritime heath and moor. Farther north still, of course, Arctic Terns breed on open tundra.

Arctic Terns return to their colonies in April and perform elegant courtship rituals both on the ground and in the air. The nest is merely a scrape in sand, but among rocks or on moorland it may be lined with a substantial amount of vegetation. Nests and eggs are effectively indistinguishable from those of Common Terns, which tend to prefer sites in longer vegetation if the two are together and which are less regularly found on open moorland.

Like other nesters on open moors, the Arctic Terns are aggressive towards people (and animals such as sheep) who walk too close to the nests. They circle above, calling repeatedly, until the intrusion becomes too much; then they hover and dive, striking with their bills and causing considerably more pain and damage than any skua, and even drawing blood.

Of all the moorland-nesting seabirds, the skuas are the most exciting. **Arctic Skuas** are slender, fast-flying and dramatic birds that breed in the Arctic regions; in Iceland, the Faeroes, Scandinavia, Scotland and along the northern rim of Siberia. They manage to breed in wet and windy, windswept places as well as areas of intense cold. Arctic Skuas typically prefer the lower slopes and drier ground, leaving higher moors and wet boggy tracts to Great Skuas. They also nest farther inland, as in the Flow Country of northern Caithness and Sutherland. They nest in colonies, but the pairs are rather loosely scattered and have relatively little in the way of concerted action or the cooperative behaviour so typical of nesting terns.

Nests are on open moorland, usually in short heather. They are established in April or May, and the early-season display by pairs and later performances by non-breeding birds are wonderful. They chase each other, tumbling in mid-air, with loud calls that enhance the wild nature of these erratic, high-speed flights. Males defend their territories by diving at rivals on the ground, but these attacks may persist for hours on end, with the grounded bird crouching at the last second, bill turned up at the attacker. If the grounded bird is equally aggressive, it will leap up at the diving male, rolling over to hit out with its feet before falling back in an undignified heap.

The two disputing skuas may opt for a long aerial pursuit, each taking turns to chase the other. One will strike out at the other, which rolls over and defends itself with its feet. This use of the feet is unusual in a species with webbed toes and no strong talons or particularly sharp claws.

Displays include roller-coaster switchback flights with vertical climbs and dives, all accompanied by ecstatic wailing calls that count among the most evocative and uplifting of all bird calls. Sometimes six or seven skuas will fly together early in the season.

Predators, innocent sheep and intruding people are watched carefully from a mound as they approach. If the intrusion proceeds, the determined bird will drop down to the ground and perform an elaborate broken-wing trick, squealing pathetically as it falls and flutters through the heather with one wing drooped and apparently useless. Should this fail to lure the potential predator away, the skua will resort to a fast dive at the head, sometimes making contact with its feet. It really is dramatic and it takes some nerve not to duck: although it is not so intimidating as the dives of the much heavier Great Skua, there is a greater chance of being hit.

Long-tailed Skuas, which breed in Scandinavia but not in Britain, use the broken-wing trick much less than the Arctic Skua and it is perhaps restricted to birds with small chicks. Dives to the head may be preceded by hovering and end with a blow with the feet, but the intensity of the dive varies greatly among individuals. Persistent intruders may be pecked on the head, and the Long-tailed Skua will even stand on a person's head in order to peck more firmly!

Great Skuas are a different proposition, being really big and fearsome creatures. They are far less Arctic in their chosen habitat than the smaller skuas, breeding in Britain and Iceland in cool, wet, windswept places with no tall vegetation. Wet, grassy or heathery moorland is typical in Scotland, frequently above the tall sea cliffs of Orkney and Shetland. Drier heather is used by nesting Arctic Skuas and wetter grass by Great Skuas, but there is some overlap and Greats show a tendency to prefer high ridges and broad slopes rather than really wet hollows.

Great Skuas gather in big flocks, or 'clubs', at freshwater lochs and also alongside carcases of sheep or cattle, where they look almost like gatherings of vultures at a kill. They often mix with Great Black-backed Gulls in these gruesome surroundings. On the territories, the most frequent display involves raising the head and bill and holding the open wings high over the back, displaying large white primary patches. There is a good deal of silent aggressive behaviour on the ground, without the sensational sounds and acrobatics of Arctic Skuas, but there are dives and pursuits in territorial disputes. Much of the courtship behaviour also takes place on the ground, involving ritualized walks and postures and insignificant calls.

From about a week before the eggs are laid human intruders are greeted by a Great Skua hovering, or peering from a vantage point on the ground. Once eggs are laid, this inquisitiveness turns more easily to aggression, and about ten days after laying dives are at their greatest intensity. The sitting bird will give an alarm call but its mate is first to attack, diving from about 30 m to gain speed, levelling out at chest or head height, then coming in fast and direct before veering up at the last second, often hitting out with its feet. Blows are less frequent than with Arctic Skuas, but the effect is much more intimidating. If there is a persistent movement towards the nest, the bird no longer circles away and dives in again; but simply sweeps up, stalls in a wing-over turn and dives back immediately. Sometimes a pair will attack in concert, one coming in from behind as the observer is intent on watching the progress of the other bird in front: not playing the game!

Any passer by will know when he has entered a skua colony: Great Skuas fly in low and swoop up at the last second, while Arctic Skuas (above) often dive from a height and strike the intruder on the head.

Island-top nesters

Linking the birds of the cliffs with those of the moor above, there are certain species that ideally like to nest on sloping or flatter ground on islands. They will never be far from the sea, so are not true moor-nesters in the sense that Common and Lesser Black-backed Gulls and Arctic Skuas are, but neither will they nest on the open ledges of a bare rock face.

Some, such as gulls and Fulmars, nest out in the open, rarely more than shaded by an overhang or sheltered by a clump of vegetation. Others, such as shearwaters, nest in burrows or deep inside the dark recesses of fallen boulders and broken rock.

Fulmars, as we have already seen, nest on all kinds of islands, and regularly occupy sites that would seem far from ideal if the population is high. Then they can be found on almost level ground or on the merest apology for a ledge, where weathered earth has slipped from a bank to form a bare hollow. Their relatives, the shearwaters and storm-petrels, are, however, more truly birds of the island slopes and ridges.

The **Manx Shearwater** is abundant on some islands off the west coasts of Britain and Ireland, with concentrations off western Irish headlands and on the islands of south-west Wales and the Inner Hebrides. On Rhum they breed nearly 3 km inland at more than 600 m above the sea, but that is rather exceptional. Skokholm and Skomer are more typical sites, if home to particularly large colonies. Here the ground is hummocky and riddled with holes, there being many thousands of Puffins, many more thousands of rabbits and tens of thousands of shearwaters all living in burrows. The tunnels are partly hidden by a thick growth of Sea Campion and, in parts, dense, aromatic Bluebells. It is essential, for their own safety and the well-being of the birds, that visitors walk only on marked and well-worn paths and do not venture over this honeycombed ground, where burrows are likely to collapse beneath their feet.

The shearwaters return to land only at night and then prefer the dark nights when there is no moon. This is because they are pretty helpless when grounded and need to avoid running the gauntlet of the Lesser and Great Black-backed Gulls, which enjoy nothing more than a shearwater as an evening meal.

When the shearwaters are arriving at the colony, the air is filled by the sound of thousands of birds making the most riotous noise, with insane screams and cackles. Evidently, to the shearwaters, these are individually recognizable and help each bird to find its mate inside a tunnel in the dark, which is itself a remarkable feat. The calls are, even to our ears, endowed with recognizable and repeatable patterns, with something like 'it-i-corrrk-A' being a typical call. A grounded bird shuffles on its belly, its wings partly open, and it uses wings, feet and bill to gain height on a slope or eminence of some kind in order to take off again.

European Storm-petrels nest in smaller colonies, sometimes only a handful of pairs loosely associated at some rocky prominence or in the broken stone walls of some ancient building. They prefer small islands, often the most isolated ones, nesting in short tunnels, crevices and hidden hollows between fallen boulders. They like storm beaches

Burrows in turf, cracks in walls or cavities in rocks above a clifftop provide the Storm-petrel with a suitable nest. The birds return to their nests only after dark, for their own safety.

clear of the highest tides, cliff faces and rocky outcrops, often above the cliffs where the land levels out except for odd, isolated rocky knolls. In Ireland, they nest in the beehive cells of old monasteries;on Shetland, in the stone walls of great brochs. Side tunnels in rabbit or shearwater burrows may be made by the petrels themselves.

Like the shearwaters, they come to land at night. They are much quieter on the wing, except in courtship chases when they churr and chirrup, but their long, purring and hiccuping sounds come from the burrows where an unpaired bird wishes to attract a mate or a paired bird perhaps calls to its mate before the eggs are laid. Who really knows what level of communication by sound, as well as orientation using smell, goes on in a shearwater and storm-petrel colony by night?

Again like the shearwaters, the storm-petrels are helpless on land, scuttling out from their burrows on their bellies. Once clear, they use their wings to flutter along, rising up on their toes. They rarely stay long on open ground: photographs of birds conveniently sitting beside burrows are, generally, 'staged', with a petrel hoiked out of its burrow to have its picture taken. Unlike Manx Shearwaters, however, storm-petrels are likely to be seen on moonlit nights, when they may fly around the breeding site in fluttering displays.

The nest is a mere depression at the end of the tunnel or cavity. If an earth burrow is used, the nest scrape may be lined with bits of wool, feathers, stones and grass stems, forming a thick pad, but rock nests are generally left unlined. A single egg is laid in June, incubated by both sexes in turns of two or three days. The egg will take around forty days to hatch, and it is a further sixty or seventy days before the chick flies.

Leach's Storm-petrel is a much rarer bird than the European Storm-petrel, its colonies far fewer and more remote. In Britain there are only about five colonies, and only four of them are regular and substantial (on St Kilda, the Flannan Isles, North Rona and Sula Sgeir; a fifth was recently discovered on the Shetlands, with endoscopes used to see into the depths of occupied burrows).

These are remote rocky sites but the nests themselves are in places with deeper soil than is often the case with European Storm-petrels. In such cases the Leach's digs its own burrow, but crevices in rocks and holes in cliffs are also used.

Strictly nocturnal above ground when at the colony, Leach's Storm-petrels are difficult to study and are among the more romantic and mysterious of our seabirds. They make a wider selection of calls than European Storm-petrels, with a typically high or lower, guttural variation on harsh, laughing sounds uttered in flight, while a long, crooning churr and strange screams come from the burrows. If birds can be made out against the sky, their long-winged shape helps distinguish them from the smaller, stubbier European Storm-petrels. Chicks in the nest make loud peeping calls, which allow some of the burrows to be detected later in the season.

Herring Gulls do not much mind if they are on a cliff face or a flat platform above, but they tend to a greater fondness for broken ground and steeper slopes than is the case with the Lesser Black-backed Gull. In many places the two are mixed together, but the preferences are still often quite distinct. On south-western islands, the broad ridges and flatter expanses away from the cliffs have mainly Lesser Black-backs while the clifftops have the Herring Gulls.

Herring Gulls are abundant and successful and can only be admired for their opportunism and fortitude. They do, however, often create problems for other species, especially terns, which frequently arrive at the colony a little later in the spring to find the best ground already taken by the more aggressive, dominant gulls. Puffins suffer similarly on certain islands, but the smaller birds may be helped by human intervention, if gull control is carefully and properly exercised.

Herring Gulls return to their colonies in March, and the males are aggressive defenders of their chosen territories. If their ritual postures, which are designed to assert dominance without coming to blows, fail, then the gulls will fight, frequently with vicious-looking holds with the bill and apparent ferocity, although lasting damage is rare.

Colonies may number as many as several hundred pairs, so there is always plenty of noise and energetic activity. Upright postures denote aggression, with the more taut, strained postures indicating greater degrees of aggressiveness. The head is raised, the bill pointed downwards, and the wings are held away from the body. Compared with a similar but inquisitive, rather than aggressive, pose, the eyes are slightly closed, the head feathers less sleek, and the wings more drooped. Greater aggression is indicated when the bird begins to move, in a stiff-legged walk, towards its neighbour. One or both birds then pull up large tufts of grass in their bills but, if there is no obvious submission from its opponent, the aggressor will suddenly charge forwards, raising its carpal joints as it does so; then it stops and fully extends its wings.

There are other displays centred around calls made in special postures. The 'long call', typical of several large gulls, involves a deep forward bow with bill open and the beginnings of the call followed by a sudden downward movement of the head before this is swung forwards and held up, bill wide open, during the classic screaming, laughing notes which everyone recognizes as the 'seagull' cry.

Courtship involves a number of stereotyped actions. A pair will strut side by side, each with its head pushed forwards and held low, the bill open and pointing slightly downwards, while they give soft mewing or moaning calls. They also perform 'choking' actions together at the likely nest site, standing closely side to side, wingtips crossed above their raised tails, head and neck hunched and bill, slightly opened, pointing down at the ground. Females also preview the postures taken up by juveniles later, begging for food from their mates with legs bent, head hunched down and bill held upwards and open as they call.

Feeding of females by males is common among seabirds in the period leading up to egg-laying. It may be a way for the females to test the qualities of the males – good males find plenty of food (and can therefore feed the family later); but it is also a way for the female to remain at or near the nest, protecting the site while using up little energy, and while gaining extra food which helps build up the large, protein-rich eggs.

Also frequent among seabirds, especially the gulls, are postures which indicate that the bird is not going to be aggressive: a deliberate submission, or at least a pose that shows the bird is not hostile. These include hunching into the submissive, food-begging posture and 'facing away', in which the bird turns its head, removing the bill, its main weapon, from view and revealing the undefended back of the head.

All these (and several other) postures combined with appropriate calls confer a high degree of order on the colony. A noisy rabble to the uneducated eye, the noisy gulls in a big colony are communicating very well. A sudden disturbance may break the rhythms and barriers within the gull colony, for a time, but the combination of threats, fights, submissions and restoration of territorial rights, all performed through ritual and symbolism, soon returns the birds to order.

Lesser Black-backed Gulls look much like Herring Gulls and are clearly closely related. In fact, as both tend to change in size, shape and coloration geographically, it is difficult, in some parts of the world, to know whether the local gull is one or the other. In the UK it is clear that they are separate and they look about as different as they ever do, but they are nevertheless close in behaviour and will even, very rarely, hybridize.

Audouin's Gulls are rather special. They are rare and restricted in range, hardly Atlantic birds at all, being essentially a Mediterranean species. Most nest on islands, although some are on little more than sandbanks at the mouth of mainland estuaries. Typical rocky-island sites are rarely more than 50 m or so above the blue Mediterranean waves, and the gulls choose exposed places on outer islands, free from disturbance. Fortunately, numbers have increased with protection in recent years and the species, which might have been on the brink of sinking into a seriously endangered state, is now apparently secure. Some of the larger colonies, though, have recently run into trouble because of overfishing of their preferred prey by humans.

The aggressive postures of Audouin's Gulls are much as those of the Herring Gull, with the upright posture (head and neck upright, bill tilted down) being characteristic, but Audouin's tends to hold the bend of the wings (carpal joints) further out from the body, while the tips remain tight over the tail. Many of the actions recall those of the Common Gull in detail, more than those of the Herring Gull.

Roseate Terns are birds of rocky islands. They sometimes nest, as, for instance, in the Azores, on dramatic, steep islets and stacks under tall sea cliffs, but other sites are lower and lie off sandy shores and dunes. Many sites have undergone a serious (in some cases, it seems, terminal) decline, and the species is now globally endangered and easily the rarest breeding seabird of the North Atlantic. Even in the Azores, apparently remote sites are sometimes visited by yachts and a casual afternoon picnic has more than once caused the complete failure of a season's breeding efforts.

A speciality of the Mediterranean Sea, Audouin's Gull is still one of the world's rarest gulls despite a welcome increase in numbers in recent years. It has a characteristic pale grey bloom on the underparts, leaving only the head and neck really pure white in summer.

While Common Terns, with their slightly longer legs than Arctics, choose relatively longer vegetation and Arctics (with tiny legs) nest in barer places, Roseates seem to prefer rocks with crevices deeply and thickly overgrown with tall, shrubby plants such as Tree

Studies of Roseate Terns may have come too late to save them in Britain, although the situation in Ireland and the Azores shows more cause for optimism. The reasons for their disastrous decline over recent decades are still not clear.

Mallow. They nest right down in the shade and shelter of the vegetation, often quite out of sight from a distance. On sandier islands they choose nest sites under overhanging Marram and Lyme Grass. The nest scrape itself may be more thickly lined with dried vegetation than those of other terns. Some colonies have been helped by the provision of nestboxes, which helps reduce predation. In mixed terneries, Roseates can be picked out by their whiter underparts and paler upperparts than Common or Arctic Terns and by their distinctive calls: one a rasping sound, and the other a squeaky double whistle.

Of the auks, only the **Puffin** is likely on the open top of an island. The others prefer rock ledges and crevices on cliffs. Puffins are much more mobile on the ground, waddling about freely on their short legs (whereas a Guillemot does little more than sit upright, resting on the full length of its legs and not standing up on its toes).

Puffin islands are among the best seabird islands of the British Isles. They tend to be in the west (with a few notable exceptions, such as the Isle of May), where cool winds bring frequent rain and mist but also many glorious, blue days with clear skies and an atmosphere so clear that one can see for many miles across the ocean. The islands themselves are green-topped, often ringed with Red Sandstone cliffs which are fringed with the white and pink of campion and thrift, and dotted with the familiar shapes of Herring and Great Black-backed Gulls and the squat forms of incubating Fulmars. Nothing could present a fresher, cleaner, wilder face.

The Puffin's large, colourful bill is essentially a summer ornament, as the Puffin at sea in winter is a much duller creature, its outer bill-casings shed to leave a small, greyish triangle, without the lines of blue, red and yellow or the fleshy yellow rosette at the gape. Spring Puffins also develop immaculate, soft grey facial patches and strange red and grey strips above and below the eye. Clearly, to a Puffin, the facial appearance is all important in the breeding season. The ridges and stripes on the bill can be used to tell the bird's age. These decorations, however, have little value in the business of catching and carrying fish and are unnecessary in winter.

The chief call of the Puffin is a deep, slightly quacking, long-drawn note that conveys a curiously contented, comfortable impression, although it may be used in aggressive encounters and take on a much angrier tone. Puffins often fight, thrashing their wings and rolling down slopes, bills clasped together in a tight, unyielding grip until they sometimes literally roll off the edge of a cliff and have to fly apart to avoid being killed on the rocks below.

They are incurably curious, nosy birds. A pair which is engaged in intimate courtship always attracts the attention of others, which gather around to watch the fun and attempt to join in. They roll and waddle, bob their heads and tap their bills. The billing is a major feature of courtship, in which the birds of a pair vigorously rub their bills together. This stimulates others to do the same until a whole sub-group of a large colony may take up the bill-clashing and bill-rubbing actions.

Early arrivals at a Puffin colony gather on the sea. This may happen in March or, farther north, early April. It may be some two weeks before any of the birds actually move up to the land. Even then, bad weather may cause a desertion for several days and the process of rather gingerly edging up to the top of the island may have to begin all over again.

The first landfall of large flocks is exciting. Suddenly the numbers on the sea increase and the Puffins become restless. They take off and fly around in whirling clouds, then stream down to the land, most of them making landfall for the first time in seven months. If the flock

No seabird sight is more endearing and entertaining than a large colony of Puffins. So often set amongst tussocks of pink thrift, the gatherings are also colourful and capture the essence of the fresh sea air and dazzling light above an Atlantic cliff.

is disturbed, they will all fly off again and whirl around some more before coming back, if at all. They may not return for some days.

These mass flights are repeated during the season at intervals, and it is then that the true numbers of Puffins present at a large colony can really be appreciated. On St Kilda, numbers were once said to be so large that photographers had to adjust their exposures when the

flocks took flight, literally darkening the sky. Most colonies hold a few hundred birds, the better ones a few thousands, and some have tens or hundreds of thousands.

The Puffins are hole-dwellers, like shearwaters, but they are in and out by day, not restricted to dark night-time visits. The best sites for Puffins are burrows dug into deep, soft soil at the top of a cliff slope or on the rounded top of an island, but many, especially on mainland sites, nest in fans of debris on cliff faces and in deep crevices leading from large cliff ledges where Razorbills and Guillemots nest.

The burrow may be a side branch from the tunnel used by a Manx Shearwater or a rabbit. Old burrows are often cleared out in spring. A Puffin will walk up to the entrance of a burrow, which may be choked with vegetation and fallen debris, and bend down to peer inside. Then it walks purposefully into the hole and spurts of earth fly out as it clears the burrow with its feet, which are equipped with curved, needle-sharp claws for the job. Inevitably, this brings other Puffins walking in to take a closer look and fights may break out, or bouts of billing, until it seems difficult to see how any pairs of Puffins ever get anything done.

They line the nest with bits of grass or feathers, in a haphazard sort of way. This, too, is often interrupted by squabbles and fights over desirable feathers or grass stems. Puffins are always undeniably fun to watch. Once the chicks are hatched, there is endless activity with Puffins flying in and out, carrying Sandeels and other tiny fish in their unique way, several together held crosswise in the bill. Puffins are frequently harried by large gulls and skuas and are regularly killed by Great Black-backed Gulls and Peregrines.

Eventually the chick is deserted and will stand at the end of the burrow on late July evenings before making its way, alone, to the sea. This is a real contrast with the young Guillemot, which is tempted down to the sea by its father long before it can fly and is attended by its parents for some weeks afterwards. The young Puffin is alone in the world from the start. Many wait until dusk before making the final move from the burrow, which may involve a walk of several hundred metres to find a suitable spot from which to launch themselves towards the sea.

Dune- and beach-nesters

Seabirds that choose to breed on flat beaches, or on sand-dunes just inland from broad sand or shingle beaches at the mouth of estuaries, face several special problems.

Those species that habitually nest close to the water's edge face the dangers of spring tides, which wash away many eggs of terns, especially Little Terns. Birds on sandy beaches risk their eggs and chicks being swamped by driven sand in summer storms. All of them are likely to face problems of disturbance by people. On sandy beaches and dunes, a couple of bathers might keep a whole colony of terns off their eggs all day long. It takes just one insensitive walker at the wrong moment to cause a Sandwich Tern colony to desert, or to put birds off eggs long enough for gulls and crows to nip in and eat their fill. Now, it is just as likely that an off-road vehicle will tear up and down the beach creating total havoc. Even helicopters have been known to land in a tern colony, with disastrous results.

Natural mammalian predators also find ground-nesting birds on beaches irresistible. Hedgehogs are great egg-thieves. Foxes eat many eggs and chicks, and can return night after night to harvest adult terns on eggs. In some places mink and Otters feed voraciously on adult birds, eggs and chicks. Rats, and cats, are frequent problems, too, even on islands.

Indeed, one wonders why some species of seabirds persist in nesting on such dangerous ground at all. They are immensely frustrating to conservationists. Most, however, unlike beach-nesting waders such as Oystercatchers and Ringed Plovers, are colonial and this, at least, offers some degree of protection from predators.

Sandwich Terns are handsome, quarrelsome, meddlesome birds of the beach. They are biggish terns, a size up from Common and Arctic Terns and more powerful in proportion. They are restricted to the coast in Britain, but a few breed inland in Ireland, beside large loughs. Most colonies are of substantial size and the nests are packed in as dense as they can be. Incubating birds are little more than 30 cm apart, just about the sitting bird's limit for a good, sharp peck at the neighbour without having to rise off the eggs.

Sand-dunes, often with good, clumpy growths of sharp Marram Grass; shingle beaches; a few artificial islands set in shallow coastal lagoons; and well-vegetated rocky islets are typical colony sites. Sandwich Terns are the earliest terns to arrive in Britain, returning in March and April and quickly occupying the colony. Colonies are erratic, however, some being occupied for years and then suddenly deserted, and others in full swing in the early spring only to empty overnight as the birds move elsewhere. Such mobility seems to be in response to disturbance, predators and the quality of the fishing, and it allows whole colonies to move on from a risky site to some-where better, with a greater chance of success.

It is clear that some pairs are formed, or reinforced, on migration or even in winter quarters in West Africa, but others are not cemented until the arrival at the colony in spring. Then there is a good deal of courtship and display, with pairs, threes and fours flying up to a great height and then returning in long, descending glides with constant calls.

All the sea terns have black caps, but that of the Sandwich Tern is emphasized by an extravagant sweep backwards from the nape into a spiky crest. In aggressive encounters on the ground, which are frequent, the crest is raised and spread, although it is the black frontal area of the forehead that is perhaps most dominant in such behaviour. Certainly it is the forehead that soon turns white in midsummer, reducing the aggression and stimulus of the black cap of spring, perhaps allowing the serious business of feeding the young to be continued without constant interruptions and squabbles.

Pairs are very demonstrative, greeting each other with open bills, or facing each other with heads raised, wings drooped and bills sharply angled down to the ground. The larger colonies are frequently separated into closely synchronized sub-groups. These may form as clusters of terns settle down at intervals, separated perhaps by a day or two of bad weather which interrupts nesting. In undulating dunes, large colonies may actually be a series of such sub-groups that happen to use suitable nesting areas close together. Broader areas of more uniform terrain allow the groups to become a single, unbroken colony, although the timing of nesting remains slightly out of phase from sub-group to sub-group.

Black-headed Gulls are quite frequently found breeding beside Sandwich Terns, and both species may derive some benefits from the

In a dense colony, Sandwich Terns space themselves out to about the maximum distance that can be defended by a stab of the bill without leaving the nest and eggs.

association. Black-headeds may steal food from Sandwich Terns (and may perhaps even benefit from the sharp eyes and quick reactions of their nervous neighbours), but the terns gain a degree of protection from the larger gulls should a predator approach. This relationship, however, is far from clear, as gull colonies may move on when terns arrive and terns with no gulls close at hand seem equally able to deal with predators on their own. Sandwich Terns are, after all, quite formidable with their determined dives and long, sharp bills. They rarely strike, but they still dive at human intruders just as Common and Arctic Terns do.

Common Terns are generally the most familiar terns, at least in southern and central Britain and much of Ireland, although Arctic Terns become more frequent in the far north. Common Terns nest in a wide variety of sites, from artificial rafts on the Norfolk Broads and gravelly islands within flooded aggregate quarries to well-vegetated saltmarsh, open dunes, shingle beaches and even banks of empty cockle shells piled up at the end of a headland by hundreds of years of tidal action.

They are noisy, irritable creatures, alert and full of life and fight, always ready to take offence and start a panic or a concerted attack on an innocent intruder. A more deadly foe, such as a fox, may be driven off by the synchronized action of the colony (which is one of the prime benefits of colonial nesting), although even a hedgehog or a few rats may be much more deadly if they choose to attack at night. In the darkness, the terns somehow lose their nerve and prefer discretion rather than valour, making off until the raid is over. By then, all too often, the damage is done.

Many pairs arrive together in spring, clearly having cemented their bond on the migratory journey from West Africa. It is possible that some pairs stay together all winter. Others, though, appear to return to the same nest and meet again almost by chance rather than by design. Who knows? Do they recognize each other, even feel a sense of relief that last year's successful partner is here again and ready to make another go of it? Whatever, there will be a long period of courtship, renewing acquaintances and strengthening the partnership that is so vital to success in rearing more young.

Again, sub-groups in the colony may be closely synchronized but somewhat out of step with other groups. The need for synchrony seems to be a strong and important one. Old, successful pairs get the best nests near the 'city centre', while young birds, new pairs with little experience, are pushed to the outer suburbs where they have to take on more interference from gulls, crows, raiding foxes and even mink and perhaps nosy non-breeding terns that want to disrupt the 'calm' of colony life. The transition from a 'beginner' at the edge to an 'old hand' near the better-protected centre needs further study as

most pairs also show a tendency to return to the same nest site as before, which would logically mean that they stick at the edge and never progress: clearly they do, eventually.

A feature of Common Tern society is an active aerial expression of courtship intentions and aggression towards intruders. More than with most terns, disputes are settled in the air, with long chases and spiralling, upward flights involving territorial males. Courtship on the ground, with particular postures (such as 'bent' and 'erect'), is reflected by aerial versions (now 'aerial bent' and 'straight' postures). Pairs fly together in long climbs and sweeping, fast descents with quick transitions from one pose to another as they fly line-astern, then pass side by side and often glide down in a sideslipping, synchronized display. This has been neatly described using the comparison of a pair of hands, held out with fingers spread, palm downwards, one 4–5 cm above the other, moved in synchrony and twisted from side to side while staying the same distance apart. So the pair of terns slips and slides through the air with an unequalled grace and delicacy.

Many Common Terns will be calling on the ground, too, as they voice their recognition of mates overhead or simply seek to lure potential mates to land alongside them. Later, chicks call out to returning adults, long before one might think they could identify their parents, but the parent is also calling as it flies in with a fish. There is a great deal of individual recognition and communication going on, all the time.

Black-headed Gulls and Sandwich Terns often nest in mixed colonies, the constant bickering between them being more than made up for by the mutual benefits of increased safety in numbers.

A strange and largely unexplained pattern is the 'dread', when a tern will suddenly give an alarm call, which abruptly silences the rest of the colony, or perhaps only the birds in the immediate sub-group, and all take off out to sea, low over the water. Gradually they trickle back. Chance observations of passing predators, or very high-flying birds of prey hardly visible to the naked eye, are occasionally used to explain such dreads, but more often there is no apparent reason and the cause remains unclear.

The terns will roost nearby and visit the colony area for a few days before finally coming to land and probably spending a night or two in the colony site before getting down to courtship proper and nest-building. The nest is often little more than a scrape in soft sand or loose pebbles, made by a bird breasting the ground and turning in a circle as it kicks back with its feet. There may be a better nest with some lining (bits of marram, leaves of purslane or samphire, or even a collection of little coloured shells) and some are quite substantial, especially if built up in a cleft in some hard rocks.

The pair will defend the immediate surroundings of the nest as a tiny nesting territory, but often one or both are away feeding and they defend a special feeding territory, too, at least temporarily.

The female will often sit about waiting to be fed by the male, clearly deciding that the hard work of her mate is a better option than using up her own energy, which is needed to create the large eggs. Males may be thought to 'do very little' in the early stages of nesting compared with the egg-laying females, but they are often exerting themselves far more.

The Common Tern's eggs are like those of the Arctic Tern, and the chicks are hard to separate (whereas young Roseates have a characteristic 'spiky' look that helps identify them in a mixed colony). The male terns feed their mates before egg-laying, to help build up essential nutrients (it has been shown that more courtship-feeding has a beneficial effect on egg production and hatching success). In this courtship-feeding stage and later, when feeding the chicks, the males often eat only the smallest fish they catch and take the better ones to the nest.

Three eggs are regularly laid, but often only two hatch and perhaps only one young is fledged. The latest egg is usually smaller than the other two and less viable, either not hatching or producing a smaller chick. If the others survive, the last chick may die. If, however, one of the larger eggs or stronger chicks is taken by a predator (quite a reasonable possibility if, on average, the predator goes for the biggest meal), the later, smaller chick may find itself getting more food and 'catches up'. It seems that the third egg (and thus chick) is a 'stand-by' in case of damage to the others and can be brought on by the extra food available if demands elsewhere are reduced. Tern productivity is all about efficiency and logic, without much room for sentiment.

Right at the water's edge on a sand or shingle beach is the place to look for a **Little Tern** colony. These tiny terns are exasperating. The beaches they choose are the very ones that people prefer on a hot summer weekend. The terns can settle, lay eggs and begin to incubate steadily, then along comes a fine day and the whole year's nesting effort can be ruined. More than that, a spell of bad weather can be equally destructive, with nests washed away or buried in sand. High tides can, of course, be predicted, and sometimes tern nests have been saved by wardens carefully lifting them on to raised boxes; but this is hardly likely to work with large numbers, year after year. Sometimes you wonder why the Little Tern evolved at all, and how it ever got this far.

Some Little Tern colonies have faced desperate troubles from Kestrels, which eat the chicks. Once a pair of Kestrels has found such easy pickings, the complete chick population may go over a few days. In at least one instance, the provision of dead mice as alternative food has relieved the pressure and the terns have done very well. The fact remains, however, that most substantial Little Tern colonies (and several of the very small ones) rely almost entirely, nowadays, on wardening, whether to deter predators or to fence off the beach and ask holidaymakers to keep to a safe distance. That they are no longer the rarest British terns is to do less with an increase in their numbers than with a drastic decline in the numbers of Roseate Terns, which are heading desperately close to extinction in Britain.

Little Terns nest in vulnerable sites just above the edge of the tideline, where holidaymakers, foxes, strong winds and high tides are all potential enemies.

Meanwhile, the Little Terns themselves are wonderful creatures. Author Simon Barnes has described them as 'how you would expect a Black-headed Gull to look after it has gone to heaven', a gull which has been angelified, made smaller, slimmer, sleeker, and more delicate altogether. The voice, he admits, leaves a little to be desired, being not much more than a harsh, fast, chattering squall. Little Terns behave much as the others but their actions are so much quicker, although not necessarily more elegant or graceful. They are too light, too jerky and too fast to be quite so beautiful as a Common or Arctic Tern or so flowing as the larger gulls. Their direct flight is fast, and the hover a whirr that a hummingbird might almost be pleased with.

Little Terns, for all their tiny stature, have the characteristic verve and disrespect for humans of most terns. Sadly, although they will fly at a man's head, they will not sit on their eggs when he approaches: aggression is no substitute for calm, peaceful disinterest.

Marsh-nesters

Not all seabirds live beside the sea in summer. At least, they are not on sheer cliffs or the edges of beaches, where they are swept by salt spray and need only raise their wings to be carried off across the waves over the open ocean. Some move a little more inland, or into sheltered parts of estuaries, to nest on salt, brackish or fresh marshes.

These are, however, rather marginal in their classification as seabirds at all, at least in summer. **Black-headed Gulls** can spend much of their lives far from the sea, wintering inland and breeding on hill lakes or lagoons within lowland reedbeds, not necessarily near the coast. Some of the largest colonies are in such situations, in fens and reedswamps where pools of open water or gleaming fleets penetrate the marsh vegetation and allow gulls to nest in irregular lines around their edges. Other Black-headed Gull colonies, however, are on coastal salt marshes, sometimes mixed with terns but often almost pure except, perhaps, for the occasional pair of Mediterranean Gulls that may be tempted to settle with them.

These coastal colonies on the edge of large estuaries can number several thousand pairs, especially on the south coast of England. The gulls build quite substantial nests of grasses and other vegetation and scraps of all kinds. They are noisy, strikingly attractive in immaculate summer plumage and extremely eyecatching. There is no point in trying to hide in such exposed places, so these birds are dazzlingly white and constantly vocal. Their babble begins before dawn and goes on until dark. In the early days of the breeding season, pairs are forming and defending their tiny territories with a variety of special postures and sounds that convey aggression, submission, readiness to mate, or a desire to be fed.

The Black-headed Gull's handsome chocolate-brown hood, tilted forward so that the back of the neck remains white, is of crucial importance in these displays. In spring it is the forehead that is last to turn dark as the head feathers moult progressively from back to front, but in late summer the frontal feathers are the first to be shed, so the winter white is maintained around the bill and face for as long as possible. It seems that a wholly dark face is such a powerful force that it is retained only for the most vital part of the breeding season. It is an aggressive, dominant feature. If a gull wishes to appease an opponent, it turns its head away, in a significant movement that exposes the vulnerable nape and reveals a large patch of submissive white.

The young Black-headed Gulls are strange, gawky creatures at first, rather long-legged and long in the neck, and quite brown. They are almost more like some strange kind of wader than a gull. As they become full grown, they take on a more contrasted appearance, with much white, silver and rich tawny-brown. Their wings are still short and rounded when they make their first flight. On the water or on the ground they appear dark, but as they fly they suddenly reveal a striking pattern of white and black on the wings and show their pure white underparts. For a few weeks they have a shawl of tawny-brown, but the brown feathers are quickly replaced by white on the neck and grey on the back, and the rich colours soon fade.

In a mixed flock, Mediterranean Gulls (right and bottom) can be picked out from Black-headed Gulls by their pure white underwings and frosty, pearly-white upperwing tips.

Mediterranean Gulls breed in a strangely irregular, patchy and erratic pattern around parts of the North Sea and Mediterranean (most in the east and around the Black Sea). In Britain, they have colonized with some success but with little sign of any dramatic rise in numbers, so still only a handful of pairs nests in any year. Most are in south-east England, breeding on gravel pits or salt marshes among colonies of Black-headed Gulls. A pair hatched out a chick on the north Norfolk coast in 1992, but it soon disappeared (probably inside a Lesser Black-backed Gull).

Rather curiously, while Black-headed Gulls have brown heads, Mediterraneans actually have jet-black hoods in summer, contrasting with vivid scarlet bills and set off by thick white 'eyelids'. They are remarkably handsome birds, the more so when they take off and reveal ghostly, silvery upperwings and pure white underwings, very different from the grey of the Black-headeds. Like the latter, their displays revolve around the dramatic, contrasting hood, which is shown off in special slow, flappy flights with head drooped or extended and in strained, head-stretching postures on the ground. The feathers of the hood, especially the lower edge, can be raised to make the whole thing look larger and thus more impressive, while it can also be tilted forwards by feather movements, minimizing the extent of black at the back. The submissive white hindneck again comes into play.

Little Gulls also have black hoods, but no white eyelid marks and no contrast between head and bill. Their wings are pale grey above, edged with white, but almost black below with the same broad white trailing edge and rim around the tip. These small, tern-like gulls nest on freshwater pools and floods but are exceedingly rare breeders in western Europe.

Likewise, **Gull-billed Terns** are very rare in the north, being effectively Mediterranean birds, although a few pairs breed irregularly close to the North Sea. Gull-billed Terns do migrate over the sea and feed close to the coast, but they are really not seabirds in any strict sense, feeding mostly on flying insects caught over fields, marshes and, increasingly, rice paddies. They are rather like Sandwich Terns but lack the spiky crest and have thicker, but pointed, all-black bills and a slightly more bull-necked shape. Their long, pointed wings tend to be straighter and give a more regular, somewhat more gull-like action even when the terns are beating into a wind and dipping sideways to take their prey.

Black Terns nest in shallow water, building on floating vegetation, and are therefore, like Little Gulls, marsh-nesters which move to the sea after breeding. They are, however, much more widespread in western Europe, but they do not breed in Britain. Whiskered Terns are also 'marsh terns', but more southerly. Both species enliven a marsh with their erratic presence, flitting low across the water and marshland vegetation, dipping to pick insects from the surface or catching them in mid-air. They have a lightness of flight and a delicacy of touch that even the sea terns rarely equal. Black Terns, especially, being that little bit smaller and daintier, have a particular elegance as they feed in flight. If large numbers feed over a lake they tend to be widely dispersed, spread over the whole of the lake, but if they are disturbed, perhaps by a passing bird of prey, they may fly together in a tight flock and make off at speed with deep, scything wingbeats, like a compact flock of medium-sized waders.

BREEDING BIOLOGY

Seabirds have taken to nesting in a whole variety of sites, as we have seen, from sheer cliffs to flat-topped stacks, sandy beaches and damp, oozing marshes surrounded by glutinous mud.

Around them, they find their food in the ocean, whether it be small fish close to the shore, eaten by terns and Black Guillemots, large fish out to sea, caught by Gannets, or smaller items such as the tiny plankton and bits of floating material picked up by fluttering storm-petrels and dumpy Little Auks.

The food they eat and its abundance clearly influence their ability to breed and the number of young they can rear. No seabird, however, does the same as a Blue Tit on land, hatching out a large clutch of eggs to coincide with a sudden abundance of tiny caterpillars. Nor do they adopt the Blackbird approach, rearing the same number of young, but in three smaller broods spread out through the summer, supported by a less abundant but more reliable supply of earthworms.

With seabirds, even an abundance of food tends to support large numbers of pairs each rearing one or two young, rather than allowing each pair to produce a bumper crop. Some species manage to find food because they are able to forage over great distances, in some cases often staying away for some days between visits to the hungry chick. Large colonies of seabirds live beside productive waters (as in northern Scotland), while others live in restricted breeding sites in poorer waters and move greater distances to feed (as occurs in the tropics).

Seabirds' breeding seasons are generally related to the supply of food available to them. **Gannets** need many Mackerel to feed large, fast-growing chicks, and the Mackerel are most readily available from June to September. There are local differences, however, and even in Britain some colonies have a tendency to be earlier or later than others: Ailsa Craig, west of the waist of Scotland, has Gannets breeding a couple of weeks later than those on the Bass Rock, just to the east, while St Kilda **Puffins** lay a few weeks later than those on the Isle of May. Bryan Nelson points out, when explaining these differences, that feeding the young at the nest is but one consideration: the young also have to survive after they leave the land and head out to sea in their first autumn. There is no point taking all summer to rear a chick if it flies off, to fend for itself, inexperienced and hardly fit, only to find a lack of fish and roaring winter gales.

In the north, where there are marked seasons (and winter can be very cold, very rough and has long, black nights), seabirds nest in summer, but start as early as they can so that the young have a good chance of survival before the short days and increased stresses of winter close in. There is not time to produce a second clutch: it is strictly a one-brood season.

Seabird breeding seasons are mostly long. Incubation periods are often remarkably long in relation to the size of the bird and the egg, especially in the storm-petrels and shearwaters (the tiny egg of the **European Storm-petrel** requires an inordinate 41 days of incubation, and the bird itself, as has been noted before, is actually a few degrees colder than most).

Then the fledging period is lengthy, too, with some chicks often having to withstand periods of meagre food supplies. **Gannets**, with abundant, large and nutritious food close to hand, can supply their large, ravenous chicks and give them a good start in life much more quickly than their tropical relatives, the boobies (which struggle with much less abundant food). Nevertheless, a young Gannet is still at the nest for thirteen weeks. **Cormorants** fledge in seven weeks, but smaller **Shags** take nearly eight weeks, so the length of time required is not solely related to size. Yet these species do manage to rear several young, perhaps three or four at a time, because they hunt their food within a few minutes' flight of the colony and can return with it several times a day. New, inland Cormorant colonies may rear great numbers of young, sustained largely by the surplus trout from large reservoir put-and-take fisheries.

Manx Shearwaters go far afield to catch their food, and their chicks face a period of seventy days or more, as often as not going hungry, in their dank, dark burrows before they risk making a move. They must put on a lot of fat and eventually may be twice the weight of their parents before they fly: it is a long process. **Storm-petrels**, so much smaller, require the same period of time. Chicks are ever hungry, but require most food as they are growing their feathers and may need less just before they are about to leave the nest – the young Gannet refuses food at the end of its fledging period.

Some seabird chicks can go without food for many days, even weeks. Like swifts, which have good spells and bad spells of feeding, according to the weather, and whose young can be semitorpid for a few days to help them survive the lean spells, seabirds such as storm-petrels live an erratic life and their chicks must be able to withstand the difficult times and take advantage of the occasional glut. Young shearwaters may be fed once or twice a day but miss days at times.

Young Gannets get fed a couple of times a day, Puffins three or four times if conditions are good, while Black Guillemots may pop out of the inshore waters and feed their chicks ten or even twenty times a day. Terns feed their chicks practically hourly.

None of the tubenoses – the storm-petrels, shearwaters and Fulmar – feeds its young after the latter have left the nest. Some seabirds do, but related species differ in this respect. Guillemots, of course, jump from their cliffs to go out to sea, accompanied by their fathers, when only eighteen days old, but Puffins leave for the sea alone, at night, at an age of seven weeks. The Puffin instantly needs to look after and feed itself, while the Guillemot is protected by 'Dad' and fed by its doting parent for many weeks.

Terns and gulls, which rear larger families, feed their young long after they have left the colony, and Cormorants and Shags feed their offspring for some weeks after the nest has been deserted. **Common Terns** have even been watched apparently dropping live fish into the water in front of their young, giving them a chance to practise the fishing techniques that are all important to them in later life.

Most seabirds the world over lay a single egg. Some gulls and terns, cormorants and shags lay two or three, or even four. Birds that feed far from the nest tend to lay smaller clutches than those that can find their needs close to home. If you cannot find food easily or quickly, it is best to have only the one young but hungry mouth the feed. If a storm-petrel returned with a poor ration of food, it simply would not work to have two or three chicks: by the time it returned from a second foray, the chicks that did not eat the first time would be half-starved and the one that did would be fitter and more likely to take the next lot of food, anyway.

Even **Gannets**, which can feed easily, are too big to rear more than one enormous chick properly. The young Gannet has to be big and fat to survive after it leaves its nest, and a pair of Gannets simply could not manage to rear two such gargantuan offspring. In some species, it has been shown that older, more experienced birds lay earlier, in better sites (often, in colonies, near the middle rather than the edge, or, on cliffs, in sheltered rather than exposed places), and lay larger, more viable eggs than younger birds. They will have chosen older, better, fitter mates to feed them before egg-laying (and then to feed the growing family). In species which lay more than one egg, these prime females will lay the largest clutches, and they probably rear the most young.

With **Herring Gulls**, as with Common Terns, the third egg is the smallest and most likely to fail. With Herring Gulls it is the most likely to be preyed on, although in the Common Tern at least it seems that the small, third egg may be an 'insurance' against the loss of the earlier, bigger eggs. Larger clutches are proportionately more successful than smaller ones: the relationship is not simply a direct one, but the bigger the clutch, the better the chance of any particular egg being hatched and any particular chick being reared, up to a point. Fascinating research has shown that the overall bulk of eggs laid by a female declines on average as the season progresses: in **Kittiwakes**, later clutches are smaller than earlier ones; in Shags, the clutch size remains the same but the eggs actually decrease in size, so it comes to the same thing. Kittiwakes laying late seem to be younger, smaller, less fit birds than those laying early, which are in the peak of condition.

Of course, all this variation reflects different ways of doing the same thing in the end. Each pair of birds has just about to replace itself with viable birds (which can reach breeding age) during the lifetimes of the pair if the whole population is to remain stable. It would take very little variation from the average, necessary scale of productivity to see the population go into a catastrophic decline or whirl away in a roaring population explosion. Generally, neither extreme case occurs.

Not all eggs hatch. **Gannets** hatch nine out of ten, and other species fewer – European Storm-petrels maybe only half of the eggs they lay. Not all chicks that hatch survive until they are able to fly. Gannets, again, achieve a high average of nine out of ten, as do Shags and Manx Shearwaters, while Fulmars manage only six to eight out of ten. Small storm-petrels may again achieve only a 50 per cent success.

Gulls and, especially, the more volatile terns have a very varied success rate from year to year. In good years, with little disturbance, no predation, plenty of food and good weather, they do very well indeed, but it takes only an intruder at the wrong moment, a very high tide, a storm, or a lack of fish for the whole colony's effort to be nullified for a year. Predation, starvation and bad weather account for most losses. Chicks are more vulnerable than eggs to predation, but starvation is generally the exception rather than a regular cause of death. In recent years, overfishing (it would seem) of small fish such as Sandeels in the North Sea and northwards to the Barents Sea has caused dramatic losses and declines in numbers of Arctic Terns, Kittiwakes, skuas and Guillemots. Catches of birds in gill nets also account for a chronic drain of remarkable numbers during the winter months.

Given the chances of failure and the hard life that even a successfully fledged chick faces before it is old enough to breed, it is clear that most seabirds have to live a long adult life. They need many bites at the cherry to give themselves a good enough chance of rearing a replacement set of two. Once the birds are mature, they seem to have about equal chances of surviving to try again, whatever their age. The high death rate comes in the years before maturity, when the birds are simply inexperienced. They are not so good at locating food, and then not so adept at catching it. Nor are they so quick to escape a predator, or even to recognize one. Birds with a couple of years safely behind them, however, are, in most species, as likely to survive as older adults.

Oddly enough – although obviously, if you think about it – the Gannet and other high-achievers (with high hatching and fledging success) have the highest mortality rates in their first year of life. In the Gannet, seven out of ten die before they are a year old. As many Manx Shearwaters, however, die in their first year, too. They then need to live longer to compensate for their poorer productivity in adulthood. Only 10 per cent of Gannets survive long enough to breed, but 16 or 18 per cent or so of Razorbills and Puffins and 20 per cent of Guillemots and large gulls do so. The species which lay single eggs do, of course, need to survive much longer than those producing multi-egg clutches, in order to produce their surviving descendants.

This is not the place to go deeply into the subtleties of breeding strategies and demographic considerations, but these brief glimpses of the variety of approaches adopted by seabirds, towards getting the same essential job done, reveal how wonderful and beautiful are the natural patterns that lie behind our seabirds' behaviour.

Seabird-watching in Europe

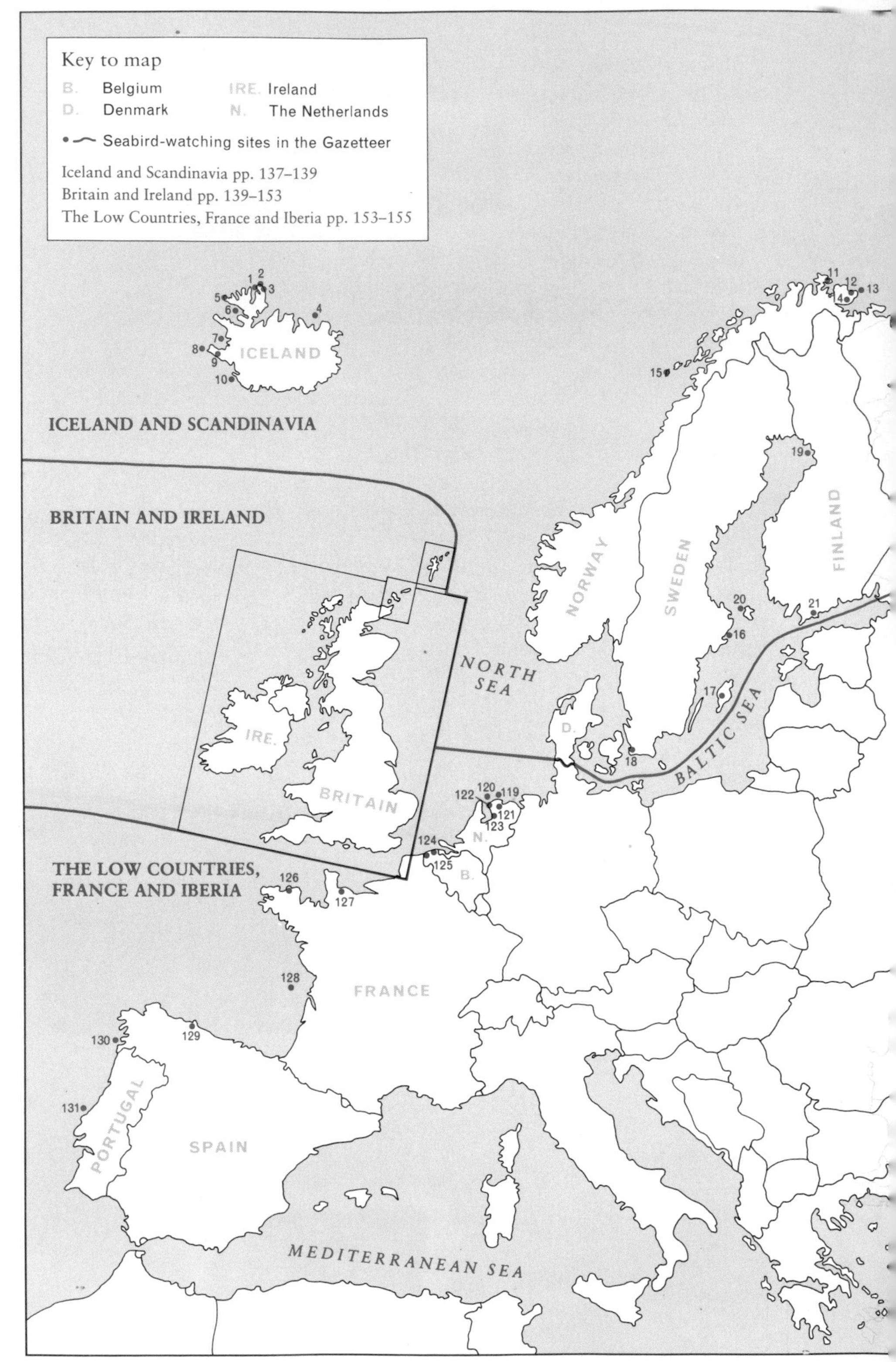

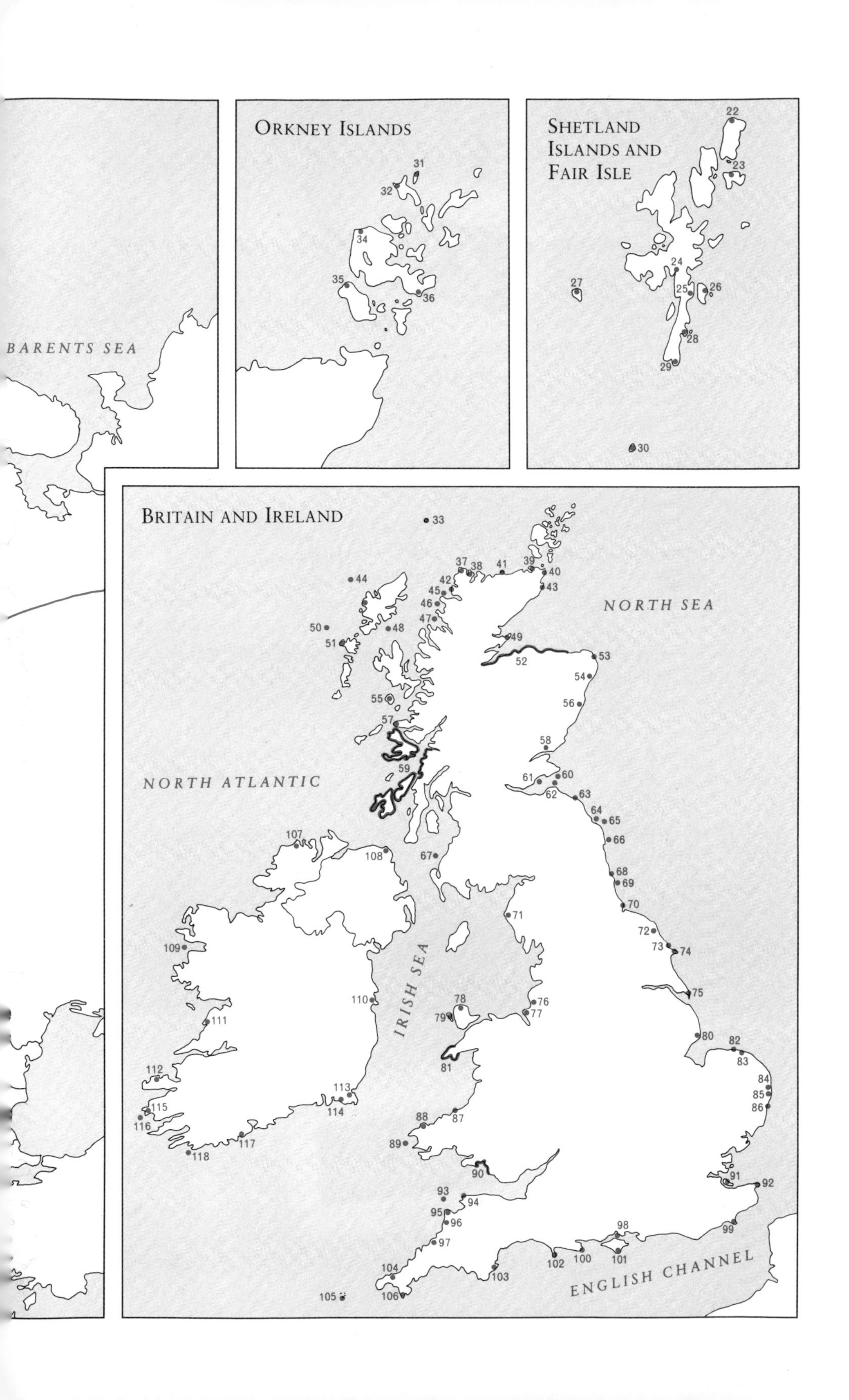

BARENTS SEA
ORKNEY ISLANDS
31
32
34
35
36
SHETLAND ISLANDS AND FAIR ISLE
22
23
24
25
26
27
28
29
30
BRITAIN AND IRELAND
33
37
38
41
39
40
43
42
45
46
47
44
48
50
51
49
52
53
54
56
55
57
58
59
60
61
62
63
64
65
66
67
68
69
70
71
72
73
74
75
76
77
78
79
80
81
82
83
84
85
86
87
88
89
90
91
92
93
94
95
96
97
98
99
100
101
102
103
104
105
106
107
108
109
110
111
112
113
114
115
116
117
118
NORTH SEA
NORTH ATLANTIC
IRISH SEA
ENGLISH CHANNEL

Gazetteer

Good places for seabird-watching fall naturally into two categories: breeding sites and migration watchpoints. The former have birds on cliffs, with many on the water below and others flying in and out, so that they can be watched at quite close range for lengthy periods. The latter are quite different, often depending on particular conditions to bring birds close enough to be seen well, and usually giving just one chance to see, identify and enjoy a bird as it flies by.

A few fortunate areas fall into both categories. Others are, in a sense, similarly well blessed in that birds on cliffs are only part of the attraction. Hundreds of others pass by offshore on the way to and from feeding areas. In that way, there is plenty of good birdwatching to be had looking out to sea, with the birds behaving as if moving by a migration watchpoint, offering plenty of experience of seabirds in action over the sea on long-distance flights, while at the same time there are others to study close at hand ashore.

Breeding sites are full of action, noise and endless fascination. Nothing beats a big seabird colony. A large, mixed colony on a big cliff above glorious, heaving seas and crashing waves is a great joy. Fulmars sail by at head height while others spar and cackle together on ledges almost within touching distance. There may be Puffins on the highest ledges and grassy slopes, with Guillemots arranged in long, dense lines on horizontal ledges, like Starlings roosting on a town hall, highly vocal and constantly active. Kittiwakes send their cries echoing around the inlets and deep, dark chasms in the cliffs. Shags brave the roughest, most irresistible waters, where the sea turns glaucous and swells enormously before bursting into foam and sparkling spray, making rainbows in the sun.

At the same time, a patrolling skua or two causes panic offshore, while Gannets catch the light, gleam spotlessly white, plunge and send up jets of spray as they hit the water headlong. A stream of Fulmars, Kittiwakes, auks and Gannets passes by over the blue sea, joined now and then by a skua or a group of tilting, flickering shearwaters. A Peregrine may chase Rock Doves or pigeons about the cliff face, Rock Pipits parachute down to the rocks below and Linnets or Twites dance along the clifftop in search of seeds.

Seals bob about in the waves or haul out on to rocks, occasionally curling the fingers of a flipper to scratch in a disarmingly, sometimes disconcertingly, human fashion. Even an occasional Minke Whale or a group of dolphins will go by or the great fin of a Basking Shark will break the surface not far out.

This gazetteer makes no claim to be comprehensive but does include a selection of interesting and exciting sites. It is arranged from north to south in three sections corresponding to the maps on pages 134–5: Iceland and Scandinavia; Britain and Ireland; and the Low Countries, France and Iberia. The bias towards Britain and Ireland is in deference to the majority of readers of this book and also because these islands are of such outstanding importance for their seabirds. The number preceding each entry refers to its location on the map.

Adult (bottom left) and juvenile (bottom right) Kittiwakes are familiar to most seabird watchers in western Europe, but Ross's Gulls (top) are rare and exciting vagrants. They breed in remote parts of the Arctic tundra and are rarely seen south of the Arctic Circle.

ICELAND AND SCANDINAVIA

Iceland

This exciting island has some outstanding seabird sites. It is of importance to the birdwatcher in the breeding season, when some of the seabird colonies are breathtaking in terms of size and settings.

1 Ritur This is a cliff at the extreme north-west tip of Iceland with 50,000 pairs of auks and many other seabirds.

2 Haelavikurbjarg and **3 Hornbjarg** Two separate cliffs on the north-west corner, with 600,000 pairs and 400,000 pairs of auks respectively. Other seabirds are also abundant, in an area of great scenic value.

4 Grimsey Although somewhat disturbed by tourists (so visiting birdwatchers should set a better example), this small island is good for birds; it lies well to the north of the Iceland mainland. It has particularly large numbers of Kittiwakes and 100,000 pairs of auks; there may just be a chance of a Little Auk here.

5 Latrabjarg A sea cliff up to 440 m high on the western promontory of north-west Iceland, where an astonishing million pairs of auks breed. There are 100,000 pairs of Fulmars (sensational enough), 400,000 pairs of Guillemots, 150,000 pairs of Brünnich's Guillemots, and a mind-blowing 250,000 pairs of Razorbills, reckoned to be over half the total world population! Puffins number 100,000 pairs and

Kittiwakes 50,000 pairs, in a simply gigantic colony that makes most seabird sites look relatively insignificant.

6 Breidafjördhur This area of relatively sheltered sea on the west has innumerable islands and many cliffs, where the visitor can find some superb seabird sites. There are many Fulmars, large Cormorant and Shag colonies, 3000 pairs of Glaucous Gulls, Kittiwakes, Black Guillemots and Puffins. Nearby Reykjavík has many Glaucous Gulls along its beaches, especially around the fish quays, which are also well worth a look for rare species in winter.

7 Hjorsey-Straumfjördhur A mixture of islands and marshes in a large, shallow bay has breeding Puffins and Arctic Terns.

8 Eldey This small island off south-west Iceland has a colony of 14,000 pairs of Gannets and a variety of other seabirds including Kittiwakes, Guillemots and Brünnich's Guillemots.

9 Krisuvik Huge sea cliffs in the south-west, where 30,000 pairs of Kittiwakes breed beside 40,000 pairs of auks, which include Puffins, Brünnich's Guillemots, Black Guillemots and many Razorbills.

10 Westmann Islands (Vestmannaeyjar) A rocky archipelago with sheer cliffs, where there are many Fulmars, breeding Manx Shearwaters and European Storm-petrels, thousands of pairs of Leach's Storm-petrels, 9000 pairs of Gannets, and many auks, including some Brünnich's Guillemots.

Norway

Norway has a long, indented western coast and 53,789 islands! It is not full of seabirds, but some sites are of immense value. Sadly, many large colonies have declined because of gill-netting and overfishing.

11 Omgangstauran Right at the very northern tip of the country, cliffs here hold 74,000 pairs of Kittiwakes, many Cormorants and small numbers of auks.

12 Syltefjordstauran In the far north in Finnmark, these coastal cliffs have the world's most northerly Gannet colony (a small one), and 140,000 pairs of Kittiwakes (Norway's largest colony), 12,000 pairs of Guillemots and as well as Brünnich's Guillemots and Razorbills. It is a stretch of coast well worth visiting.

13 Hornøy An island at the north-east corner of Norway where Shags, very many Herring Gulls, Kittiwakes, Guillemots, a few Brünnich's Guillemots and Puffins breed. This area is always likely to produce exciting seabirds offshore.

14 Varangerfjord This is a famous bird area at the far north-east of Norway. Within it there are several Arctic Tern colonies and large Kittiwake colonies at Ekkerøy.

15 Vaerøy and Røst Of several seabird islands down the west coast, these are perhaps the most important, with many Kittiwakes and great numbers of Puffins (70,000 on Vaerøy and 700,000 pairs on Røst), and large but declining numbers of other auks: both are difficult to reach.

Sweden

Sweden has relatively few major seabird breeding sites, but does have exciting migration watch points and is excellent for gulls and terns.

16 Gotland Some islands, especially off the south-western tip, have seabird colonies, with some auks and many terns.

17 Stockholm The archipelago off Stockholm has a few breeding birds, including terns and many Black Guillemots.

18 Falsterbo At the southernmost tip of Sweden, this famous migration site can provide exciting seabird viewing between the bird-of-prey and songbird movements.

Finland

There are limited opportunities for seabird-watching in Finland, but some specialities are attractive, nevertheless.

19 Krunnit Nature Reserve An island group near the head of the Baltic, notable for a hundred pairs of Caspian Terns.

Caspian Terns are the world's biggest: not only are they heavy and broad-winged, but they have extra-large bills, too. At a distance they look perfectly proportioned and clearly more tern-like.

20 Signilskär West of the small archipelago at the south-western tip, these fifty barren islands have breeding gulls, terns, Black Guillemots and Razorbills. Other islands to the south have a few breeding Caspian Terns.

21 Gulf of Finland Several national parks and sanctuaries are known for great numbers of winter wildfowl, but the whole area is superb for breeding terns, especially Arctics, and passage skuas. It has breeding Common, Herring, Lesser Black-backed and Great Black-backed Gulls, Guillemots, Razorbills and Black Guillemots.

Denmark

The indented coastline and innumerable islands of Denmark, with many shallow coastal lagoons and extensive, secluded beaches, are ideal breeding areas for terns and gulls, with smaller numbers of auks (although Black Guillemots are widespread). Sandwich, Little, Common and Arctic Terns all have important populations, although all are scattered and few colonies are large. Some freshwater lagoons hold large numbers of breeding Cormorants, and both Cormorant and Shag are common and widely distributed around the coasts.

BRITAIN AND IRELAND

Britain

With their western position, surrounded by productive seas, and with their great variety of coastlines, Britain and Ireland are of major significance for their seabird populations.

22 Hermaness Other than Muckle Flugga one can hardly go farther north

Much stiffer-winged than gulls, the Fulmar revels in a good blow, using the strong winds and fickle air currents above breaking waves to power its gliding flight.

than Hermaness, on Unst, Shetland. It is one of many Shetland seabird colonies, but many say that it is the finest cliff scenery anywhere and the birds are terrific. This is a great place for Gannets (9900 pairs), with, in recent years, a single Black-browed Albatross in summer. It also has many of the more widespread seabirds, including Puffins (28,000 pairs), Fulmars (25,000 pairs), Guillemots (21,000 birds) and Great Skuas (nine hundred pairs).

23 Fetlar has fine seabirds, and Manx Shearwaters can be seen at dusk in its broad bays. Northern stacks and headlands have Puffins, and the moors are full of Great and Arctic Skuas (220 and 143 pairs, respectively). Over 13,000 pairs of Fulmars breed, from cliffs to rooftops and low banks. Contact the RSPB warden.

24 Scalloway and **25 Lerwick** harbours are good for Black Guillemots and close views of skuas and, in winter, excellent places to look for Glaucous and Iceland Gulls, if not something even rarer.

26 Noss and Bressay Take a boat trip from Lerwick around Noss, east of Bressay, to see Gannets (7200 pairs) and vast numbers of Guillemots (37,700) from sea-level, a memorable experience. You can also get the ferry from Lerwick to Bressay and cross to Noss if it is calm, but check the times with the Tourist Information people in Lerwick. On foot, Noss is superb, but watch the skuas! There are some four hundred pairs of Great Skuas here.

27 Foula, west of Mainland Shetland, is a fine seabird island if you can get there, with immense numbers of skuas (Great Skuas total 2340 pairs, the world's biggest colony). Foula also has 46,800 pairs of Fulmars, 2400 pairs of Shags, 37,000 Guillemots and 48,000 pairs of Puffins. There are weekly flights and twice-weekly boats come from Walls, which must be booked in advance, but no day visits.

28 Mousa Mousa is a small island reached by boat from Sandwick; European Storm-petrels breed in the Pictish broch tower and on the storm beach. Make arrangements to stay late to see them.

29 Sumburgh Head Shelter in the lee of the lighthouse wall and watch Kittiwakes Fulmars, Shags and Puffins. Seawatching is good in autumn.

30 Fair Isle, that pre-eminent migration station, is noted for seabirds in summer, including a few Gannets and many Great and Arctic Skuas. Fulmars (27,000 pairs), Puffins (20,000) and Guillemots (32,000) are its most abundant birds.

The Orkneys are, if anything, even better, with colonies of Arctic Terns (especially on Westray and **31 Papa Westray** in the north, which have almost twenty seabird species breeding) and a scattering of Great and Arctic Skua colonies. Most islands have seabird cliffs: **32 Noup Cliffs**, on Westray, south of the lighthouse on Noup Head, are absolutely magnificent with 40,000 Guillemots and 25,000 Kittiwakes.

33 North Rona and Sula Sgeir, remote but exciting seabird islands north of the Outer Isles are difficult to get to but well worth it. There are 43,000 Guillemots. If you have the chance of a seabird-island cruise, take it: see the various birdwatching magazines for occasional advertisements.

34 Marwick Head is an RSPB reserve, where you will find huge numbers of auks and Kittiwakes.

On Hoy, the cliffs of the RSPB reserve of **35 North Hoy**, north from Rora Head, are excellent seabird stations, with an awesome 35,000 pairs of Fulmars and 1230 pairs of Great Skuas pre-eminent.

36 Copinsay, reached by boat from Mainland, has superb seabird cliffs.

The north coast of Scotland is blessed with beautiful cliffs, full of seabirds. Head for **37 Cape Wrath** and the giant cliffs of Clo Mor, where there is a large Puffin colony. Breeding birds here include 9000 pairs of Kittiwakes, 12,000 Guillemots and 1600 Razorbills. Seawatching can be excellent, but is neither easy nor convenient.

38 Faraid Head, north of Durness, is a small but excellent headland, with good seawatching and close views of Puffins, Fulmars and other species.

39 Dunnet Head is a fine site, with many Puffins and 15,000 pairs of Kittiwakes. The Caithness cliffs between them have 30,000 pairs of Fulmars, 13,000 pairs of Herring Gulls, 2000 pairs of Shags, 46,000 pairs of Kittiwakes, 130,000 adult Guillemots, 850 pairs of Great Black-backed Gulls, 16,000 Razorbills and 1850 Black Guillemots.

40 Duncansby Head is brilliant. The Stacks of Duncansby have grand scenery and good birds.

41 Strathy Point has few breeding birds but a constant variety of movement offshore.

42 Handa Island This beautiful isle is south of Cape Wrath, near Tarbet and Loch Laxford. Every seabird enthusiast should, one day, visit it. Boatmen from Tarbet will take you across or sail around under the towering cliffs of the Great Stack.

All the auks, Fulmars, and both skuas breed on this magical island, with its mixture of Old Red Sandstone cliffs and lovely sandy beaches. Some of its impressive bird numbers are 10,000 pairs of Kittiwakes, 98,000 adult Guillemots, 16,400 Razorbills and eighty pairs of Great Skuas. When I used to visit in the late 1960s, Great Skuas were just about colonizing.

43 Noss Head, near Wick, Caithness, is reached on a minor road north of Wick Airport. It is a small but attractive headland, with typical close views of colonial seabirds. Wick itself may have northern gulls in winter. Berriedale and Ord Point near Helmsdale have fine seabird cliffs. Brora and Golspie often have Fulmars over the town streets and terns offshore.

44 The Flannan Islands are splendidly evocative, with 22,000 Guillemots.

45 Point of Stoer and **46 Cluas Deas**, north of Stoer, and the low, rocky headland near Reiff, west of Inverpolly, are good for passing seabirds, with vast numbers of local Fulmars, Gannets and gulls and many shearwaters at times, including Sooties in late summer. These common, locally breeding seabirds are endlessly enjoyable to watch on a blowy summer day with blue skies and heaving seas. There are doubtless possibilities at many other similar sites in the area. The small north-western harbours, such as Kinlochbervie, are worth a look 'out of season' for winter visitors, especially northern gulls.

47 Priest Island, an RSPB reserve in the Summer Isles, has a huge European Storm-petrel colony of about 10,000 pairs, as well as Shags and Black Guillemots.

48 The Shiants have 18,000 Guillemots and an amazing 77,000 pairs of Puffins, one of the finest colonies in Britain.

49 Tarbat Ness, a fine, low headland stretching north-eastwards at the mouth of the Dornoch Firth, is good for watching passing skuas and shearwaters. North Sutor, to the south, at the mouth of the Cromarty Firth, has a selection of breeding birds including auks and many Fulmars.

50 North Uist: try the Balranald area, especially Aird an Runair, where the seawatching can be productive. On some spring days it is incomparable for rare skuas, and a variety of shearwaters and petrels can be expected given enough hours spent watching: as with most seawatching sites, a day visit can be disappointing. Farther north, the Butt of Lewis has a few breeding birds, but is best known as a place to watch the endless procession of Gannets from Sula Sgeir, Fulmars, Kittiwakes and auks, in a setting of real wilderness qualities, magical and timeless.

The variety of ferry services and summer pleasure-boat trips around the Minches is ideal for watching the regular breeding seabirds, with a chance of Leach's Storm-petrels and, in late summer, Sooty Shearwaters. The area is also worth exploring in winter, especially for Glaucous and Iceland Gulls. At any time, watch trawlers offshore, as they will frequently be surrounded by great numbers of gulls, including very many Great Black-backs, with Fulmars, Gannets and skuas.

51 St Kilda is a special case: hard to reach but a phenomenal seabird complex, with great numbers of Gannets, Puffins and both Leach's and European Storm-petrels. It lies well to the west of Harris. Its huge cliffs – the highest in Britain – and the great, conical stacks covered with Gannets make it unique, one of the great seabird sites of the northern hemisphere. Numbers include 63,000 pairs of Fulmars (just imagine that!), 50,000 pairs of Gannets, 22,000 Guillemots and 155,000 pairs of Puffins.

52 Moray Firth The bays and headlands of the southern shore of the Moray Firth, such as Findhorn and Burghead, similarly attract terns, Kittiwakes and attendant skuas. Farther east, Troup Head and Pennan Head is a genuine seabird-cliff complex with 16,000 pairs of Kittiwakes and 16,000 individual Guillemots. There is also a very new, small but increasing gannetry here. In suitable winds, especially northerly and easterly gales, Chanonry Point and Fort

George in the inner Moray Firth can be excellent, with skuas attacking the many Common, Arctic and Sandwich Terns there.

53 **Rattray Head**, at the end of a vast dune system north of Peterhead, is a great seawatching station, given northerly or north-east winds, when Peterhead itself should not be neglected. Expect Glaucous and Iceland Gulls there in winter. Just south of Peterhead, Longhaven Cliffs have a nice selection of breeding seabirds.

54 **Sands of Forvie**, by the Ythan Estuary north of Aberdeen, a famous tern colony, has many Sandwich Terns usually breeding. The whole sandy stretch of coast north of Aberdeen is good for terns and skuas in summer and gulls in autumn and winter.

55 **Rhum** is one of the most spectacular of the Inner Hebrides, with a famous colony of Manx Shearwaters on its mountains. There is a ferry service from Mallaig (check with Caledonian MacBrayne offices) or day trips from Arisaig. The shearwaters, of course, arrive only after dark and require a long, hard walk uphill, but you can see them offshore or from the mainland opposite on a late-summer afternoon, in inshore waters such as the Sound of Arisaig or off headlands of Arisaig and Morar.

56 **Fowlsheugh**, south of Aberdeen and Stonehaven, near Crawton, is an RSPB reserve of great character. It is a splendid seabird cliff of the very best kind, deeply indented, colourful, busy with birds. It has big numbers of Kittiwakes (36,650 pairs) and Guillemots (about 56,000 individuals), but also Razorbills (5800), Puffins and Fulmars in a marvellously scenic setting.

57 **Ardnamurchan Point** is a good seawatching station in summer and early autumn, with passing shearwaters and storm-petrels always possible along with the regular breeding species such as Gannets and Fulmars in large numbers. As with any of these west Scottish headlands, a good westerly wind helps bring the birds closer and give more spectacular watching.

58 **Firth of Tay** The outer parts of the Firth of Tay, especially Tentsmuir, east of Tayport, are worth exploring in summer and autumn for breeding and migrant terns and skuas. Fife Ness, to the south, is more genuinely a seabird-watching point, an excellent headland given good easterly weather, with Little Gulls a speciality and Sooty Shearwaters fairly regular.

59 **Western sea lochs** All the sheltered sea lochs of the west are good for scattered tern colonies and many Black Guillemots. The Firth of Lorne and Sound of Jura have Black Guillemots on innumerable rocky islets and, in late summer, large flocks of Guillemots and Razorbills gather to moult. Seabirds here are best seen from a ship; watching from headlands is less successful. Manx Shearwaters are regular in these waters and seawatching proper, in autumn, from the smaller inner isles and headlands of Mull, Jura and Islay has not been widely practised but could surely be productive. In winter, crossings from West Loch Tarbet to Islay are good for divers and northern gulls.

60 **The Isle of May**, at the mouth of the Firth of Forth, is a famous migration point and seabird colony, with Puffins a plus point (around 12,000 pairs).

61 **The Inner Forth islands** have small numbers of terns, with poor remnants of Roseate Tern colonies. Several Forth shore-line sites are good for migrant and post-breeding concentrations of terns, especially Musselburgh lagoons just east of Edinburgh.

62 **Bass Rock** is the pride of the Forth, reached by boat from North Berwick. This great rock is the home of a thriving and exciting colony of Gannets (21,600 pairs), among the most picturesque of all the Gannet cities around the British Isles. Boat trips run from North Berwick harbour.

63 **St Abb's Head** between Dunbar and Berwick-upon-Tweed, is a neglected migrant-watching point and wonderful seabird site, with birds breeding on particularly scenic cliffs. There are 620 pairs of Shags, 25,000 pairs of Kittiwakes and 33,000 Guillemots here. Strong northerly to easterly winds produce good movements, with occasional Sooty

There is much to learn in a seabird colony: Herring Gulls exhibit a fascinating range of postures and calls that help to maintain a strict hierarchy and social structure in a superficially chaotic scene.

Shearwaters and regular Pomarine Skuas in spring and autumn. The whole stretch of coast is worthy of exploration.

64 Holy Island and **65 the Farnes**, to the south, are justly much more famous. Holy Island is good for migrants and a few seabirds, including Fulmars. A visit to the Farnes is the very best way to get to know Arctic Terns (1990 pairs), Guillemots (12,975 pairs), Puffins (26,000 pairs) Shags (1690 pairs) and Kittiwakes (6150 pairs): at extremely close range. The terns may be closer than you would like (take a beak-proof hat!). There are also Sandwich Terns – and who knows what? This is the place where Europe's only Aleutian Tern was recorded and where a Lesser Crested Tern, which should breed in Africa or the Middle East, has several times paired with a Sandwich Tern and reared a chick. The Farnes are unique: even the Guillemots nest on flat-topped rocks instead of sheer cliffs.

66 Coquet Island lies just south, with its Puffins and terns, including a colony of Roseate Terns which is currently increasing (one of very few places that can boast that). There are boat trips from Amble, which are excellent for seeing the birds, but no landing is allowed on this RSPB reserve.

67 Ailsa Craig is a remarkable island, easily seen from the Carrick coast south of Ayr. There are boat trips from Girvan (check with the Tourist Information Centre there). It has a fine Gannet colony, as well as Puffins, Black Guillemots and the regular auks. Fishing Gannets can be widely watched all along the Ayr coast and around Arran and Kintyre.

Several sites around Tyne and Wear are good for seawatching, especially on windy autumn days with a chance of skuas and shearwaters, and during northerly gales in November when Little Auks are driven inshore. **68 Seaton Sluice** and **69 Whitburn** are good areas.

70 Hartlepool offers similar possibilities, with some autumn skua passages especially dramatic and exciting. The whole coast south from Cleveland to Robin Hood's Bay offers seawatching opportunities, with terns and skuas in spring, shearwaters in later summer and skuas again in autumn.

71 St Bees Head, the pointed tip of the Lake District bulge, is the best seabird site in the region, with a good variety of birds including Black Guillemots. To the south, Walney Island, near Barrow-in-Furness, has a huge gull colony (10,000 pairs each of Herrings and Lesser Black-backs) and in autumn, given a strong north-wester, it can be good for passing birds blown inshore, including Leach's storm-petrels.

72 Filey Brigg, south of Scarborough, is a small but useful headland with good seawatching, a little overshadowed perhaps by Flamborough Head to the south.

73 Bempton Cliffs, just north of Flamborough Head, provide an easy-to-reach mainland site for breeding Puffins (7000) and Gannets (1200 pairs), as well as great numbers of Guillemots (32,600 birds) and Kittiwakes (85,400 pairs). It is a splendid place, quite different in character from the northern isles or the great sandstone cliffs, as it is a white, chalky cliff, sheer and daunting. In summer, trips are arranged to watch the birds by boat from Bridlington; the same boat switches to skua and shearwater trips offshore in autumn. Contact the RSPB for details of these.

74 Flamborough Head is one of England's premier seabird sites, with skuas, terns, gulls, shearwaters and all kinds of rarities providing entertainment almost all year round. Much depends on the weather conditions, but with a strong onshore wind, especially a northerly gale, there can be huge numbers of birds moving offshore; often the movements are northerly in calmer weather after a storm, after birds have been blown south into the southern North Sea and are beating their way back out of the bottleneck again. Compared with Cornwall, there are fewer Sabine's Gulls and large shearwaters but far more Sooty Shearwaters and Long-tailed Skuas here and, later in the year, bigger numbers of Little Auks. Pomarine Skuas are most frequent in October, Long-tailed probably better earlier in autumn.

More southerly headlands such as **75 Spurn Head** have erratic but often good seawatching depending on conditions. There are always terns about in summer, frequently skuas, but a good onshore blow in autumn is best for a few shearwaters and storm-petrels and the odd rarity, such as a Sabine's Gull or Long-tailed Skua. In late autumn, Little Auk movements in northerly gales can be seen particularly well from the lower coasts (from higher cliff positions, they tend to be more distant and little more than tiny dots whirring over the waves).

The whole of the Liverpool Bay coast, including Blackpool beach, Southport, Formby, Hoylake, the Point of Ayr and Great Ormes Head, provides a grandstand view of Leach's Storm-petrel movements in late-autumn gales. Sometimes the tired birds, blown into the bay and swept around by violent north-westerlies, flutter across the beach or even overland, but viewing conditions can often be difficult with a roaring sea crashing over the promenades and rocky headlands.

76 Seaforth, Merseyside, is a relatively newly discovered site of major interest. Gales from the north-west bring out the best of the seawatching, especially if they coincide with early-morning high tides when the birds move in close. Wet weather is also a bonus, if you can cope with it.

77 Hilbre and the whole of the outer Dee Estuary attract big numbers of terns in autumn, including many Littles and, in the past, odd Roseates, although the latter are now sadly rare.

The northern headlands of Anglesey have odd breeding sites for seabirds, including a handful of Black Guillemots, and some, such as Point Lynas, are good seawatching sites. Anglesey has other good seabird spots. **78 Cemlyn Bay** is an important tern breeding area. **79 South Stack** has accessible Puffins as well as Fulmars, Guillemots, Razorbills, and Kittiwakes. Such opportunities, with easily seen seabirds, are rather rare in the southern half of Britain and, like Bempton in the east, South Stack is a godsend for people who cannot make the trek to Scotland every weekend.

80 Gibraltar Point can have good seawatching if conditions are right, especially in late-autumn gales.

81 The Lleyn Peninsula surely must be good for seabirds in autumn? Try it and make some discoveries.

Norfolk sites include **82 Sheringham** and **83 Cromer**, where shelters on the sea front give some cover in bad weather. In calm days in summer, Norfolk is sheer bliss for the fan of Common, Sandwich and Little Terns and an assortment of commoner gulls.

84 Great Yarmouth beach is the best Little Tern place in East Anglia. Great Yarmouth and **85 Lowestoft** are likely to have Glaucous and Iceland Gulls in winter.

86 Minsmere, the RSPB reserve in Suffolk, is excellent for Little Terns, though numbers of Common and Sandwich Terns there have recently declined.

87 Craig yr Adar, just west of New Quay, has seabirds, nicely visible in summer.

88 Strumble Head is a big headland west of Fishguard. It is a famous place for seabirds passing offshore. There will almost always be something to see, with so many seabirds breeding on the Dyfed islands, including big numbers of Gannets and Manx Shearwaters. In autumn, there are skua movements and, especially in a

A fishing Little Tern has such speed and zip that it is unlikely to be mistaken for any other. The body is relatively heavy for the slender wings; it makes up for that by a particularly quick, energetic action.

westerly or north-westerly wind (perhaps after a south-westerly blow has taken birds up into the Irish Sea) shearwaters and storm-petrels pass by in exciting numbers. This is a place where some rarities, such as Little Shearwaters, have turned up and must appear again.

89 The Pembrokeshire Islands For seabirds, South Wales is synonymous with the Dyfed islands: especially Ramsey, Skomer, Skokholm and Grassholm. All are beautiful and wonderful in their own way. Skomer is perhaps the finest, if the most popular and populous, a glorious mixture of red and grey sandstones, vivid bluebells, seas of sea campion and red campion and thrift. It has great colonies of immaculate Lesser Black-backed Gulls, Herring Gulls and, on the cliffs, Kittiwakes and auks, including Puffins. In the dark, the island erupts with the noise of tens of thousands of returning Manx Shearwaters. Skokholm is smaller, more compact, more exclusive, a superb secret shared by lucky visitors who may stay for a week. It, too, has shearwaters and European Storm-petrels breeding, as well as auks. Grassholm is altogether different, farther out, hard to reach, bleak and exposed. It is half covered with Gannets (almost 30,000 pairs), and soon there will be so many that those lucky enough to land will not be able to clamber to the top of the island to look down over the Gannet colony: it is already spilling over the crest of the ridge and down the other side. Check with the RSPB for visiting details, but do not be surprised, even if you have booked a trip, to find it cancelled at the last moment because of rough seas.

South Wales's bays and estuaries are great places for marine ducks but not so good for 'real' seabirds, apart from good numbers of gulls. **90 Swansea Bay** is famous for its regular Mediterranean Gulls and Ring-billed Gulls but, along with the Burry Inlet, also has many Kittiwakes in late summer and Little Gulls in summer and autumn. These do not demand seawatching conditions in the sense of onshore winds or gales, but you need to sort out the tides and time your visits accordingly.

Almost anywhere along the Kent coast can turn up Mediterranean Gulls and Yellow-legged Gulls, the latter especially in late summer and autumn. Surprising numbers of shearwaters, storm-petrels and skuas can be seen in times of gales along the North Kent coast, right up into the Thames, with the **91 Isle of Sheppey** perhaps the most productive. Across the estuary, even Southend Pier has established a reputation for seabirds in strong winds!

92 North Foreland is also a decent seawatching site.

93 Lundy, off the north coast of Devon, has declined in importance as a seabird breeding site, but it still has good birds and renewed its fame among birdwatchers in the 1990s with the annual appearance of a tiny ocean waif, an Ancient Murrelet from the North Pacific. There are regular boats from Bideford. In the past, summer sailings from Swansea, via Ilfracombe, were great for seeing European Storm-petrels in the Bristol Channel; the area is worth exploring by any means. Ferries to Rosslare and Cork, can be very exciting especially in late summer and autumn. Their birdwatching possibilities have been neglected, but for big shearwaters, especially, and regular European Storm-petrels, they are excellent.

The North Devon coast is interesting in a north-westerly gale, with situations such as **94 Morte Point,** west of Ilfracombe, worth a try.

The North Cornwall coast is similarly interesting and, the farther west you go, the better it gets. There is a whole string of magnificent headlands and bays, making this a wonderland of opportunities for the seabird-watcher who likes his birds in a fine setting. **95 Trevose Head, 96 Bedruthan Steps, 97 Kelsey Head** and other notable outposts have birds offshore all year.

The marshes and sand and shingle spits around **98 The Solent and Southampton Water** have numerous colonies of Black-headed Gulls and several of

Common, Little and Sandwich Terns. Mediterranean Gulls are always possible. Langstone Harbour and Chichester Harbour also have gull and tern colonies, with Little Terns carefully protected on some islands.

99 Dungeness, aside from its passage offshore, has several lagoons formed by flooding gravel pits on the huge shingle formations. Islands within the lagoons have Black-headed and occasional pairs of Mediterranean Gulls, odd Arctic and even Roseate Terns among the Common Terns and, on passage, frequent Black Terns and Little Gulls. Offshore there is a famous warm-water outflow from the power station, 'the patch', which attracts feeding gulls and terns at almost any time.

100 St Alban's Head, near Swanage, has Fulmars, Kittiwakes and Guillemots, and Durlston Head has Razorbills, too.

There are a few breeding seabirds on the Isle of Wight, and **101 St Catherine's Point** is well placed for seabird passage, especially for skuas in spring. Pomarine Skuas pass up the English Channel in early May: a good onshore wind often brings small groups close inshore. These can be watched from many Channel headlands and watching points, from Portland to Dungeness, with Beachy Head and Selsey Bill particularly favoured by seabird enthusiasts.

102 Portland Bill in Dorset can be good but can be as dead as anywhere: given good conditions it is a fine place for offshore passage (with Pomarine Skuas in spring), and there are breeding auks (including a handful of Puffins) on its western cliffs.

103 Berry Head, south of Torbay, is a boon to south-western birdwatchers in summer, because it allows them to watch commoner breeding seabirds – Guillemots, Razorbills, Kittiwakes and Fulmars, with a few Puffins – at close range on their breeding ledges.

104 St Ives Bay fills with birds that have been blown north by a south-westerly gale and turned inshore as the wind backs to the north-west. The conditions are short-lived and need quick reactions, but St Ives Island can be a sensational place to see numbers of Kittiwakes, auks, Fulmars and Gannets stream past, with a variety of shearwaters possible and at times dramatic movements of European Storm-petrels and skuas, bringing almost regular rarities with them. Great and Cory's Shearwaters, Long-tailed and Pomarine Skuas, Sabine's Gulls, even Wilson's Storm-petrels and Little Shearwaters might fly by; but most people who go yearly are hardly likely to bump into the real rarities, and there are a good many birds claimed by the optimistic and imaginative! Watch from the shelter of the wall at the coastguard station, if there is space.

All the way to Land's End there are good places to watch, but Pendeen Watch and Cape Cornwall are as good as any. Pendeen may be better than St Ives in a westerly wind; watch from near the lighthouse. Just around the corner, so to speak, Gwennap Head and Porthgwarra provide great seabird-watching at times with a more southerly wind: in late summer there can be many Cory's Shearwaters (but, regular as they may seem in the bird reports, they remain the exception rather than the rule).

105 The Isles of Scilly are best for breeding seabirds, although some good seawatching is possible in autumn gales. Roseate Terns still nest near Tresco but Common Terns are much more likely. There are a few Manx Shearwaters and European Storm-petrels. In autumn, Kittiwakes haunt the sheltered sounds, where huge rafts of Shags provide the most remarkable seabird sights in the islands.

106 The Lizard is similar to Porthgwarra, but less likely to produce the real rarity, and heading east the headlands tend to be progressively less exciting.

Ireland

The coasts of Ireland are punctuated by breathtaking seabird stations. Surrounded by seas, exposed to the roaring winds from the Atlantic, its climate ameliorated by the

Gulf Stream, Ireland has many characteristics that make it ideal for the birds and those who wish to see them.

107 Donegal headlands Malin Head, Horn Head, Rossan Point, although in some cases in the shadow of offshore islands, are worth visiting in seawatching seasons. Horn Head, near Dunfanaghy, has breeding auks, including Puffins, and perhaps had as many as 45,000 pairs of Razorbills in 1969 though recent counts have been much lower.

108 Rathlin Island is off the north Antrim coast, north of Ballycastle, from where there is a daily boat. Its excellent seabird cliffs have breeding auks and Kittiwakes, and many passing Gannets on most days. Guillemots have increased to almost 40,000 individuals, Razorbills to 7000.

County Mayo has many islands and, like Kerry, has a long list of sites of interest for seabirds. Most colonies are rather small, and you can travel around some glorious coastal countryside, drinking in the Atlantic scenery and picking up groups of terns, gulls, auks and Fulmars here and there rather than finding individual sites of high drama.

If you can get to **109 Clare Island** you will see more birds, but they are rendered

The appeal of seabirds is typified by a view of Fulmars and Gannets over a turbulent sea, against the backdrop of a thriving seabird island.

insignificant by the gigantic cliffs which rise sheer from the sea west of Clew Bay. Nearby islands, such as Caher and Inishturk, are worth exploring for storm-petrels.

The east coast has few dramatic cliffs but does have important tern sites, although many tend to be somewhat erratic. Sea-watching is also productive at times from a number of headlands, and winter gull flocks attract occasional Ring-billed and other rare gulls.

110 Sandymount Strand, Dublin, often has large tern flocks (including many Arctics) and has attracted large roosting concentrations (including up to 7000 Common Terns) in recent years, from late July to early September. Rockabill Island, off the Dublin coast, is one of Europe's best sites for Roseate Terns.

South of Galway Bay and the Aran Islands are the **111 Cliffs of Moher**, among the finest mainland cliffs in the country.

There is a visitor centre and car park on the R479 north of Liscannor, and the cliffs allow unsurpassed views of seabirds including Puffins, Razorbills, Guillemots and Fulmars. Puffins have declined drastically, however, while Guillemots have increased to around 12,000 and Razorbills to rather more than 2000.

Kerry is a splendid county of long headlands, rugged islands and deep fingers of sea penetrating far into its mountains. A list of Manx Shearwater and European Storm-petrel colonies reveals the lyrical names of forgotten islets: Inishvickillaun, Inishnabro, Inishtearaght, Inishtooskert, Skellig Michael, Illaunturlough, Gurrig. For the

The highest pinnacles on a clifftop often form the nest site of a pair of Great Black-backed Gulls. They are strikingly impressive, but powerful predators and eaters of carrion: a dead sheep may sttract these gulls as well as Ravens and Buzzards

autumn seawatcher, it is a wonderful region. Many headlands are great in a good gale, few better than **112 Brandon Point**, looking north from the Dingle peninsula south of the Shannon. In a north-wester it is, like St Ives in Cornwall, a place of mouthwatering potential. Kerry Head, Valentia and, going south and eastwards through Cork, Dursey Island, Mizen Head, Cape Clear Island and the Old Head of Kinsale are all grand sites.

113 The **Wexford lagoons** and coastal lakes are good for terns in summer and autumn, and a variety of gulls frequents the coast, with a rare one from time to time. Wexford and the coast up to Dublin is also a fine stretch for Little Gulls in winter.

114 Great Saltee can be reached by boat from Kilmore Quay near Wexford and has Manx Shearwaters and smallish numbers of Gannets nesting, as well as the more usual range of seabirds visible by day. Guillemots number around 10,000 pairs, with Razorbills rather fewer.

115 Puffin Island does indeed have Puffins, perhaps as many as 10,000 pairs, plus Fulmars, Kittiwakes, Razorbills, Guillemots and Shags. Bull and Cow Rocks and Blasket Island, off the mouth of Dingle Bay, have many breeding seabirds.

116 Great Skellig, off Valentia Island, is a seabird island with many Razorbills as well as its tubenoses. Little Skellig has a unique appeal, with jagged cliffs housing many thousands of breeding Gannets.

117 Old Head of Kinsale, south-west of Cork, is also a breeding-bird station, as are the Saltee Islands off Wexford. Wexford, indeed, is a good seabird county, despite being rather tucked away into the Irish Sea, away from the full-blown Atlantic of Kerry and western Cork.

118 Cape Clear Island has unbeatable records of shearwater and storm-petrel passages in autumn, and birds there seem to come exceptionally close to shore on a regular basis. In spring, south-westerlies and westerlies associated with warm fronts bring movements of Pomarine Skuas.

THE LOW COUNTRIES, FRANCE AND IBERIA

The Netherlands

The Netherlands is mostly low, flat and often bleak but, to the surprise of many who contend that a few hills are needed to make a place worthwhile, it is a beautiful and compelling country to visit. In winter it is full to the brim with wildfowl, but it is not a great place for seabirds.

119 Terschelling is one of the Friesland islands and has 13,000 pairs of Lesser Black-backed and 21,000 pairs of Herring Gulls. Common and Arctic Terns also breed. Little Gulls appear on passage.

120 Texel, the isle of birds, is a remarkable island with good seawatching at times which might bring a few shearwaters and auks, interesting gulls (rarely, a Sabine's) and terns at the right time of year. The ferry from the mainland often produces a few Guillemots and Razorbills. Some 12,000 pairs of Black-headed Gulls breed on its reserves, and 750 pairs of Sandwich Terns, with lesser numbers of Little, Common and Arctic Terns.

121 Lauwersmeer and the nearby harbour are worth a look for passage and wintering Little Gulls, with the odd Glaucous Gull or Iceland Gull from time to time. Black Terns are frequent in good numbers on migration.

122 Balgzand in the Waddensee is part of this terrific complex of mudflats and marshes where some terns breed, but it is remarkable chiefly for its Black Terns, which have reached 30,000 birds at autumn roosts. Gull-billed Terns are also likely here.

123 The IJsselmeer and adjacent lagoons and marshes (including Oostvardersplassen) are famous mainly for wintering wildfowl but also have passage terns in plenty. This region may have 100,000 Black Terns in autumn and there may well be 20,000 Common Terns, 2000 Little Gulls and small numbers of Caspian and Gull-billed Terns about, too.

Gulls often follow ships, riding in the slipstream to save energy while searching for food churned up in the wake. Here two Sabine's Gulls (right) have joined the more usual Herring and Lesser Black-backed Gulls.

Belgium

Belgium has no coastal cliffs and few sites which are of any real significance for breeding seabirds.

124 Het Zwin has around ten pairs of Mediterranean Gulls and 370 pairs of Common Terns. Krekengebied, just east, has breeding Black Terns in a mixture of fresh and brackish lagoons.

125 Vlaamse Banken on the coast is a breeding site for Mediterranean Gulls and has over 500 pairs of Common Terns.

France

The best parts of France for seabird watchers are in the north-west and along the Channel.

126 Les Sept-Iles in Brittany comprise several rocky islands which have an interesting and important selection of breeding seabirds. Around fifty pairs of Fulmars and, more interestingly perhaps, a few pairs of Manx Shearwaters and European Storm-petrels breed here. There is also a fine Gannet colony, with 6000 pairs, and good numbers of Shags and Cormorants, a few Kittiwakes and small numbers of Guillemots, Razorbills and Puffins. The archipelago of Moléne has more Manx Shearwaters and 300–400 pairs of European Storm-petrels. Cap Sizun may also have breeding storm-petrels, and holds a few Fulmars, good numbers of Shags and over 800 pairs of Kittiwakes. Other island groups in Brittany have small storm-petrel colonies and tern colonies, including erratic and declining clusters of Roseate Terns.

127 Falaises du Bessin That this is the most important site in France for Fulmars (74 pairs) and Kittiwakes (just 1300 pairs) shows how deprived of breeding seabirds are the French and how lucky those in northern and western areas are by comparison. This site is on the English Channel coast in Basse Normandie. A little to the west are the Isles Saint-Marcouf, where 500 pairs of Cormorants and several thousand pairs of gulls breed.

128 The Bay of Biscay is an important area for migrant seabirds, and ferry crossings in late summer and early autumn can be productive. They can also be very rough! This whole area is full of potential, with Cory's and Great Shearwaters, Wilson's Petrels and, later, Sabine's Gulls all likely.

Spain

Mainland Spain is relatively poor for seabirds, but it does have an interesting mixture of northern and southern species.

129 Cabo Torres has breeding European Storm-petrels (150 pairs) and Shags.

130 Islas Cíes in Galicia, north-west Spain, is important for breeding Yellow-legged Gulls of the Atlantic race (12,000 pairs), as well as Shags (400 pairs) and a tiny handful of Guillemots. Cabo Vilan to the north has 30–40 pairs of Guillemots, the best group remaining in Spain of the special Iberian race.

Portugal

Like Spain, Portugal has a few northern species in declining colonies, while southern specialities breed at a few sites. There is, however, much more potential for good migration watching.

131 The Berlengas Islands off the coast of Portugal, north of Lisbon, have especially interesting seabirds, if in small numbers. Here, Cory's Shearwaters and Madeiran Storm-petrels breed, the latter at their only site in Europe. Shags, Lesser Black-backed and Yellow-legged Gulls and a declining population of Guillemots nest. Sadly, the last species seems to be following the Kittiwake, which recently disappeared from the islands as a breeding bird. If anyone could watch passing seabirds from the islands or from boat trips in the region, there would surely be much to see.

BIBLIOGRAPHY

Brooke, M., *The Manx Shearwater*, T. & A.D. Poyser, London, 1990

Boag, D., and Alexander, M., *The Atlantic Puffin*, Blandford Press, Poole, 1986

Burgur, J., and Gochfeld, M., *The Common Tern*, Columbia University Press, New York, 1991

Cramp, S. (Ed.), *The Birds of the Western Palearctic*: Vol. I, Oxford University Press, 1977; Vol. III, OUP, 1983; Vol. IV, OUP, 1985

Del Hoyo, Elliott and Sargatal, *The Handbook of the Birds of the World*, Vol. 1, Lynx Ediciones, Barcelona, 1992

Delin, H., and Svensson, L., *Photographic Guide to the Birds of Britain and Europe*, Hamlyn, London, 1988

Fisher, J., *The Fulmar*, Collins, London, 1982

Flegg, J. *The Puffin*, Shire, Aylesbury, 1985

Freethy, R., *Auks: an ornithologists's guide*, Blandford Press, Poole, 1987

Furness, R.W., *The Skuas*, T. & A.D. Poyser, Calton, 1987

Gooders, J., *The New Where to Watch Birds*, Andre Deutsch, London, 1986

Grant, P.J., *Gulls, a guide to identification*, T. & A.D. Poyser, Calton, 1982

Grimmet, R.F.A., and Jones, T.A., *Important Bird Areas in Europe*, ICBP, Cambridge, 1989

Harris, M.P., *The Puffin*, T. & A.D. Poyser, Calton, 1984

Harrison, Peter, *Seabirds, an identification guide* revised edition, Christopher Helm, London, 1985; *Seabirds of the World, a photographic guide*, Christopher Helm, London, 1987

Howard, R., and Moore, A., *A Complete Checklist of the Birds of the World*, Academic Press, London, 1991

Hume, R., *The Common Tern*, Hamlyn, London, 1993

Hutchinson, C., *Birds in Ireland*, T. and A.D. Poyser, Calton, 1989

Lockley, R.M., *Ocean Wanderers*, David and Charles, Newton Abbot, 1974; *Puffins*, Dent, Letchworth, 1953

Lofgren, L. *Ocean Birds*, Croom Helm, London, 1984

Lloyd, C., Tasker, M., and Partridge, K., *The Status of Seabirds in Britain and Ireland*, T. & A.D. Poyser, London, 1991

Madders, M., and Welstead, J., *Where to Watch Birds in Scotland*, Christopher Helm, London, 1989

Marples, G. and A., *Sea Terns or Sea Swallows*, Country Life, London, 1934

Nelson, B., *The Gannet*, T. & A.D. Poyser, Berkhamsted, 1978; *Living with Seabirds*, Edinburgh University Press, 1986; *Seabirds, their biology and ecology*, Hamlyn, London, 1980

Norman, D., and Tucker, V., *Where to Watch Birds in Devon and Cornwall*, Croom Helm, London, 1984, revised edition, Christopher Helm, London, 1992

Pritchard, D.E., *et al.*, *Important Bird Areas in the United Kingdom including the Channel Islands and the Isle of Man*, RSPB/JNCC, Sandy, 1992

Saunders, D., *Where to Watch Birds in Wales*, Christopher Helm, London, 1989

Sharrock, J.T.R., *Atlas of Breeding Birds in Britain and Ireland*, British Trust for Ornithology, Tring, 1976

Thom, V.M., *Birds in Scotland*, T. & A.D. Poyser, Calton, 1986

Warham, J., *The Petrels: their ecology and breeding systems*, Academic Press, London, 1990

Scientific Names

All bird species mentioned in the text are listed below, with their scientific name, in systematic order.

Emperor Penguin	*Aptenodytes forsteri*
Jackass Penguin	*Spheniscus demersus*
Great Northern Diver	*Gavia immer*
Black-browed Albatross	*Diomedea melanophris*
Royal Albatross	*D. epomophora*
Wandering Albatross	*D. exulans*
Fulmar	*Fulmarus glacialis*
Cory's Shearwater	*Calonectris diomedea*
Great Shearwater	*Puffinus gravis*
Sooty Shearwater	*P. griseus*
Manx Shearwater	*P. puffinus*
Mediterranean Shearwater	*P. yelkouan*
Little Shearwater	*P. assimilis*
Wilson's Storm-petrel	*Oceanites oceanicus*
European Storm-petrel	*Hydrobates pelagicus*
Leach's Storm-petrel	*Oceanodroma leucorhoa*
Madeiran Storm-petrel	*O. castro*
diving-petrels	*Pelecanoides spp.*
tropicbirds	*Phaethon spp.*
boobies	*Sula spp.*
(Northern) Gannet	*Morus bassanus*
Olivaceous Cormorant	*Phalacrocorax olivaceus*
Cormorant	*P. carbo*
Double-crested Cormorant	*P. auritus*
Shag	*P. aristotelis*
Brown Pelican	*Pelecanus occidentalis*
Magnificent Frigatebird	*Fregata magnificens*
Grey Heron	*Ardea cinerea*
flamingos	*Phoenicopteridae*
Wigeon	*Anas penelope*
Tufted Duck	*Aythya fuligula*
Red Kite	*Milvus milvus*
White-tailed Eagle	*Haliaeetus albicilla*
Lammergeier	*Gypaetus barbatus*
Sparrowhawk	*Accipiter nisus*
Golden Eagle	*Aquila chrysaetos*
Osprey	*Pandion haliaetus*
Kestrel	*Falco tinnunculus*
Hobby	*F. subbuteo*
Peregrine	*F. peregrinus*
Pheasant	*Phasianus colchicus*
Coot	*Fulica atra*
Oystercatcher	*Haematopus ostralegus*
Ringed Plover	*Charadrius hiaticula*
Knot	*Calidris canutus*
Dunlin	*C. alpina*
Common Sandpiper	*Actitis hypoleucos*
Pomarine Skua	*Stercorarius pomarinus*
Arctic Skua	*S. parasiticus*
Long-tailed Skua	*S. longicaudus*
Great Skua	*S. skua*
South Polar Skua	*S. maccormicki*
Mediterranean Gull	*Larus melanocephalus*
Laughing Gull	*L. atricilla*
Franklin's Gull	*L. pipixcan*
Little Gull	*L. minutus*
Sabine's Gull	*L. sabini*
Bonaparte's Gull	*L. philadelphia*
Black-headed Gull	*L. ridibundus*
Audouin's Gull	*L. audouinii*
Ring-billed Gull	*L. delawarensis*
Common Gull	*L. canus*
Lesser Black-backed Gull	*L. fuscus*
Herring Gull	*L. argentatus*
Yellow-legged Gull	*L. cachinnans*
Iceland Gull	*L. glaucoides*
Glaucous Gull	*L. hyperboreus*
Great Black-backed Gull	*L. marinus*
Ross's Gull	*Rhodostethia rosea*
Kittiwake	*Rissa tridactyla*
Ivory Gull	*Pagophila eburnea*
Gull-billed Tern	*Gelochelidon nilotica*
Caspian Tern	*Sterna caspia*
Royal Tern	*S. maxima*
Lesser Crested Tern	*S. bengalensis*
Sandwich Tern	*S. sandvicensis*
Roseate Tern	*S. dougallii*
Common Tern	*S. hirundo*
Arctic Tern	*S. paradisaea*
Aleutian Tern	*S. aleutica*
Forster's Tern	*S. forsteri*
Bridled Tern	*S. anaethetus*
Sooty Tern	*S. fuscata*
Little Tern	*S. albifrons*
Least Tern	*S. antillarum*
Whiskered Tern	*Chlidonias hybridus*
Black Tern	*C. niger*
Black Skimmer	*Rynchops niger*
Guillemot	*Uria aalge*
Brünnich's Guillemot	*U. lomvia*
Razorbill	*Alca torda*
Great Auk	*Pinguinus impennis*
Black Guillemot	*Cepphus grylle*
Ancient Murrelet	*Synthliboramphus antiquus*
Little Auk	*Alle alle*
Puffin	*Fratercula arctica*
Rock Dove	*Columba livia*
Snowy Owl	*Nyctea scandiaca*
swifts	*Apus* spp.
bee-eaters	*Merops* spp.
Great Spotted Woodpecker	*Dendrocopos major*
Rock Pipit	*Anthus petrosus*
Dunnock	*Prunella modularis*
Blackbird	*Turdus merula*
Coal Tit	*Parus ater*
Blue Tit	*P. caeruleus*
Great Tit	*P. major*
Treecreeper	*Certhia familiaris*
Starling	*Sturnus vulgaris*
Linnet	*Carduelis cannabina*
Twite	*C. flavirostris*

INDEX

Page numbers in *italics* refer to illustrations